T0135983

Bibliografische Information der Deutschen Nationalbibliothek

Die Deutsche Nationalbibliothek verzeichnet diese Publikation in der Deutschen Nationalbibliografie; detaillierte bibliografische Daten sind im Internet über http://dnb.d-nb.de abrufbar.

ISBN 978-3-8325-2566-8

Logos Verlag Berlin GmbH
Comeniushof, Gubener Str. 47,
10243 Berlin
Tel.: +49 (0)30 42 85 10 90
Fax: +49 (0)30 42 85 10 92
INTERNET: http://www.logos-verlag.de

Processing strategies of the auditory system for improving the detection of masked signals

Von der Fakultät für Mathematik und Naturwissenschaften
der Carl von Ossietzky Universität Oldenburg
zur Erlangung des Grades und Titels eines
Doktors der Naturwissenschaften (Dr. rer. nat.)
angenommene Dissertation

Bastian Epp, M.Sc. (Eng. Phys.)
geboren am 22. Dezember 1981
in Heilbronn

Gutachter: Prof. Dr. Jesko L. Verhey
Zweitgutachter: Prof. Dr. Volker Mellert

Tag der Disputation: 31.05.2010

Wiederholen zwar kann der Verstand, was da schon gewesen; was die Natur gebaut, bauet er wählend ihr nach. Über Natur hinaus baut die Vernunft, doch nur in das Leere: - du nur, Genius, wehrst in der Natur die Natur.

(Johann Christoph Friedrich von Schiller)

Contents

1 Introduction **7**
- 1.1 General introduction 7
- 1.2 A review of the literature 10
  - 1.2.1 Processing of binaural cues 11
  - 1.2.2 Processing of across-frequency cues 13
  - 1.2.3 Processing of combined monaural and binaural cues 15
  - 1.2.4 Mechanical preprocessing at the level of the cochlea 17
- 1.3 Aim and scope of this thesis 18

2 Superposition of masking releases **21**
- 2.1 Introduction 23
- 2.2 Methods 30
  - 2.2.1 Procedure 30
  - 2.2.2 Stimulus and apparatus 30
  - 2.2.3 Subjects 34
  - 2.2.4 Model 34
- 2.3 Results 41
  - 2.3.1 CMR as a function of the IPD 41
  - 2.3.2 CMR as a function of the portion of comodulated masker components 46
- 2.4 Discussion 49
- 2.5 Summary and Conclusions 54

3 Combination of masking releases for different center frequencies and masker amplitude statistics 57
3.1 Introduction 59
3.2 General methods 63
3.2.1 Procedure 63
3.2.2 Stimuli and apparatus 64
3.2.3 Listeners 65
3.3 Experiment 1 - CMR and BMLD at 700 Hz . . . 65
3.3.1 Rationale 65
3.3.2 Methods 66
3.3.3 Results 66
3.4 Experiment 2 - CMR and BMLD for higher and lower frequencies 72
3.4.1 Rationale 72
3.4.2 Methods 72
3.4.3 Results 73
3.5 Experiment 3 - transposed stimulus 77
3.5.1 Rationale 77
3.5.2 Methods 77
3.5.3 Results 78
3.6 Discussion 81
3.6.1 Comparison with previous studies focussing on either CMR or BMLD 82
3.6.2 Comparison to previous data on the combined effect of interaural differences and comodulation 85
3.6.3 Implications for the underlying mechanism 88
3.7 Summary & Conclusions 91

4 Modeling cochlear mechanics: Interrelation between cochlea mechanics and psychoacoustics 93
4.1 Introduction 95
4.2 Model description 100
4.3 Model evaluation 105
4.3.1 Response to pure tone stimulation 106

4.4 Model application 110
4.4.1 Response to two-tone stimulation 110
4.4.2 DPOAE growth functions 115
4.4.3 Finestructure in SFOAE 117
4.4.4 Threshold in quiet 118
4.4.5 Modulation detection threshold and threshold finestructure 124
4.5 Discussion 127
4.5.1 The role of the damping profile in the nonlinear characteristics of the model 127
4.5.2 Evaluation of the nonlinearity 129
4.5.3 Reflection due to cochlea roughness . . . 130
4.5.4 Application to psychoacoustics 131
4.6 Summary & Conclusions 133
A Complexity of the damping profile 134
B Variation of the roughness 137

5 Summary and concluding remarks **139**

Bibliography **142**

List of figures **157**

List of tables **160**

Kurzfassung

Das auditorische Systems erlaubt es, komplexe akustische Szenen in einzelne Elemente, sogenannte akustische Objekte, zu unterteilen. Während beispielsweise ein normalhörender Mensch auch in einer komplexen Umgebung einzelne Schallquellen (einzelne Objekte) von den restlichen Schallquellen (dem akustischen Hintergrund) trennen kann, ist diese Fähigkeit bei Menschen mit einer Schwerhörigkeit häufig eingeschränkt. Zur Entwicklung von Hörhilfen, die diese Einschränkungen kompensieren können, ist ein detailliertes Verständnis der Verarbeitungsstrategien, also der Implementierung und Verschaltung einzelner Verarbeitungsstufen, im gesunden System notwendig.

Die vorliegende Arbeit ist in fünf Kapitel unterteilt. Im ersten Kapitel wird die Problemstellung beschrieben und in die vorhandene Literatur eingegliedert. Kapitel 2 bis 4 befassen sich mit der Untersuchung der Verarbeitungsstrategien von kombinierten frequenzübergreifenden und ohrübergreifenden Signaleigenschaften bei Normalhörenden und dem Einfluss der cochleären Vorverarbeitung auf psychoakustische Effekte. Im letzten Kapitel werden die Ergebnisse der einzelnen Studien abschließend zusammengefasst.

Kapitel 2 behandelt die Detektionsleistung normalhörender Versuchspersonen mit Hilfe eines kombinierten Paradigmas zur Untersuchung von comodulation masking release (CMR) und binaural masking level differences (BMLD). Das in der Literatur beschriebene Flankenbandparadigma zur Untersuchung von CMR wird erweitert, um die systematische Kombination von kohärenten Pegelschwankungen in verschiedenen Frequenzbereichen (Komodulation) mit interauralen Phasenunterschieden (interaural phase differences, IPD) zu untersuchen. Die Ergebnisse bei einer Mittenfrequenz von 700 Hz, bei der sowohl Komodulation als auch IPD getrennt voneinander zu einer ähnlich großen Verbesserung der Mithörschwelle führen, legen dar, dass eine Kombination von Komodulation und IPD zu einer Addition der Schwellenverbesserungen in Dezibel (dB) führt. Das entwickelte effektive Modell unterstützt die Hypothese einer seriellen und unabhängigen Verarbeitung von Einhüllendeninformationen über Frequenzen hinweg und Unterschieden in der Feinstruktur zwischen den Ohren entlang der Hörbahn.

Die Generalisierung der Ergebnisse aus Kapitel 2 folgt in Kapitel 3. Das Paradigma aus Kapitel 2 wird auf einen größeren Frequenzbereich zwischen 200 Hz und 3000 Hz erweitert. In diesem Frequenzbereich sind die Effekte durch Komodulation und IPD unterschiedlich stark ausgeprägt. Außerdem schließt das Experiment in Kapitel 3 eine weitere, in der CMR-Literatur benutzte Stimuluskondition mit ein. Zusätzlich wird die Unabhängigkeit der Verarbeitung von Einhüllendeninformationen und Feinstrukturunterschieden bei hohen Frequenzen getestet. Die Ergebnisse zeigen eine Additivität von CMR und BMLD, wenn die spektrale Bandbreite des Stimulus berücksichtigt wird. Ebenso unterstützen die Ergebnisse die Hypothese unabhängiger Verarbeitung von Einhüllendeninformationen über Frequenzen hinweg und Feinstrukturunterschieden zwischen den Ohren über einen großen Frequenzbereich.

Als Grundlage zur Abschätzung des Beitrags der cochleären Vorberarbeitung zu den psychoakustisch erfassten Größen beschreibt Kapitel 4 die Entwicklung und Anwendung eines aktiven, nichtlinearen Modells der Cochlea. Die Nichtlinearität des Modells ist im Gegensatz zu vorherigen Studien als eine doppeltsigmoide Dämpfungsfunktion implementiert. Das Modell berücksichtigt wichtige Aspekte der mechanischen Vorverarbeitung einer gesunden Cochlea. Die Anwendung des Modells zeigt, dass gleichzeitig physiologische Daten und psychoakustische Daten normalhörender Versuchspersonen nahe der Ruhehörschwelle mit Hilfe eines physikalischen Modells der Cochlea beschrieben werden können.

Die Ergebnisse der Arbeit verdeutlichen, dass das auditorische System des Menschen in der Lage ist, Komodulation und interaurale Phasenunterschiede unabhängig voneinander über einen großen Frequenzbereich zu nutzen. Die Verbesserungen der Mithörschwelle durch Komodulation und interaurale Phasenunterschiede können durch diese unabhängige Verarbeitung optimal genutzt werden. Das zeigt die gleichzeitige Nutzbarkeit frequenzübergreifender und räumlicher Information zur auditorischen Objektbildung. Die Anwendbarkeit des entwickelten Cochleamodells auf psychoakustische Fragestellungen stellt die Basis dar, cochleäre und neuronale Beiträge zu psychoakustisch gemessenen Effekten zu separieren und getrennt voneinander zu analysieren. Das hier entwickelte psychoakustische Paradigma und das hier entwickelte funktionelle Modell bieten die Grundlage für die Untersuchung verwandter Fragestellungen zur Detaillierung der Implementierung und Anordnung der Verarbeitungsstrategien mit Methoden der Elektrophysiologie oder bildgebenden Verfahren. Das physikalische Modell der Cochlea stellt die Grundlage dar, den Einfluss cochleärer Vorverarbeitung auf die akustische Wahrnehmung genauer zu untersuchen und damit die den psychoakustischen Effekten zu Grunde liegenden neuronalen Mechanismen genauer zu beschreiben als in auditorischen Modellen mit einer Filterbank als cochleäre Vorverarbeitung.

Abstract

The processing of the auditory system allows the separation of complex acoustical scenes into single elements, so-called auditory objects. While normally hearing listeners can extract single sound sources (auditory objects) from all other sound sources (the acoustical background) also in a complex acoustical environment, this ability is reduced in people with a hearing impairment. In order to develop technical devices to compensate for this impairment, a detailed knowledge about the implementation and the interrelation of the single processing stages is necessary.

This thesis is divided into five chapters. The problem statement is given in the first chapter, followed by a review of the relevant literature. Chapters 2 and 3 address the investigation of the processing of combined across-frequency and binaural signal properties in normally hearing listeners. Chapter 4 concentrates on the influence of cochlear processing on psychoacoustic effects. The results of the single studies are summarized in the last chapter.

In the study described in chapter 2 the performance of normally hearing listeners in a combined paradigm to investigate comodulation masking release (CMR) and binaural masking level difference (BMLD) is examined. The CMR flanking band paradigm as described in the literature is extended to systematically investigate the combined effect of coherent envelope fluctuations in different frequency regions (comodulation) and interaural phase differences (IPD) on masked thresholds of a tone. The results at a frequency of 700 Hz where both, comodulation and IPD separately lead to a similar decrease in masked threshold, suggest that a combination of comodulation and IPD leads to an addition of the benefits in decibels (dB). The developed effective model supports the hypothesis of a serially aligned and independent processing strategy of envelope information across frequency and finestructure differences across ears along the auditory pathway.

The generalization of the results of chapter 2 is shown in chapter 3. The experimental paradigm of chapter 2 is extended towards a frequency range between 200 Hz and 3000 Hz. In this frequency range, the effects due to comodulation and IPD have different magnitudes. In addition, another stimulus condition as used in studies in CMR was included and the independence of envelope and finestructure cues is tested at a high frequency. The results show an additivity of CMR and BMLD if the spectral contents of the stimulus is accounted for. Likewise they support the hypothesis of independent processing of across-frequency envelope and interaural finestructure cues over a broad frequency range.

Chapter 4 presents the development and the application of an active and nonlinear cochlea model to experimental paradigms with the aim to provide a basis to estimate the extent to which cochlear processing contributes to the psychoacoustically measured effects. In contrast to previous studies, the nonlinearity is implemented as a double-sigmoidal damping function. The model accounts for various aspects of the mechanical preprocessing of a healthy cochlea. The application of the model shows that physiological data and psychoacoustical data of normally hearing listeners can be accounted for.

The results of this thesis point out that the auditory system is able to process comodulation and interaural phase differences independently and over a broad frequency range. Hence, improvements of masked thresholds can be used in an optimal manner. These results show the simultaneous usability of across-frequency and spatial cues to form auditory objects. The ability of the developed cochlea model to account for psychoacoustical data provides the basis to separate peripheral contributions to psychoacoustical effects from neural contributions. The psychoacoustical paradigm and the functional model developed in the present thesis can be used to further investigate the implementation and location of the underlying neural processing mechanisms with methods of electrophysiology and imaging techniques. The physical model of the cochlea provides the basis to further quantify the influence of cochlear preprocessing on auditory perception and to describe the underlying mechanical and neural processes in more detail compared to auditory models with filter-based preprocessing approaches.

1 Introduction

1.1 General introduction

The auditory system of mammals, and especially that of humans, is able to extract relevant sounds from an acoustical background in complex listening conditions. A famous example is the ability of human listeners to listen to a single speaker during a cocktail party [Cherry, 1953]. Acoustically such a condition is very complex, as it is a mixture of many single sound sources such as speech from different persons at various spatial positions, music and a mixture of other environmental sounds. In addition, all these sounds are affected by the properties of the room where the situation is located, which are different for a small office room than for a large church. Nevertheless, a normally hearing listener is able to segregate single sound sources from this mixture of sounds, e.g., to separate a speaker from all other speakers and from other background sounds. This concept is often referred to as auditory object formation, as an analogy to object formation in the visual domain where the visual scene is grouped into objects such as, e.g., tables, chairs and people [Griffiths and Warren, 2004]. The ability to form auditory objects is not only important for humans but also for other mammals, for example in order to segregate potential predators or prey from the acoustical background composed of sounds from other animals or the environment.
To form auditory objects, the auditory system makes use of physical properties of the acoustic signals. The efficiency with

which the auditory system processes these physical properties exceeds that of all technical systems by far. This is true as long as all parts of the system are working properly. An impairment of any kind at any stage of the processing path might lead to a distinct decrease in performance, leading to the inability to attend to single speakers at a cocktail party or to detect an approaching predator.
Two of the acoustical properties which can be used by the auditory system (so-called "cues") are disparities in the signals arriving at the two ears (interaural disparities) [e.g. Licklider, 1948] and coherent envelope fluctuations in different frequency regions of the signal (comodulation) [Hall *et al.*, 1984].

The most common cases where interaural disparities arise are listening conditions where the sound emitting source is not directly in front of the listener but, for example, at a certain angle remote from the viewing direction of the listener. Due to the differences in the distance to the ears, the sound reaches the ear of the listener closer to the source before it arrives at the ear further away. This leads to interaural level differences in the arrival time between the two ears (interaural time differences, ITD) which in the case of sinusoidal signals is equivalent to interaural differences in the phase (interaural phase differences, IPD). Depending on the wavelength, the sound wave can also be attenuated by the presence of the head by diffraction of the sound wave, leading to differences in intensity between the two ears (interaural level differences, ILD). With the information provided by the interaural disparities, the position of the sound source in space can be evaluated by the auditory system. The knowledge of the position of a sound source can then in turn help to segregate this sound source from the acoustical background composed of sound sources located at other spatial locations.

In contrast to interaural disparities where the evaluation of the signal at two ears is necessary, comodulation can also be used when only one ear is stimulated. Comodulation is a common property of many natural sounds including speech [Nelken *et al.*, 1999]. A speech signal consists of many frequency components which are changed in amplitude and center frequency to form complex spectro-temporal patterns over time. Single segments of these patterns form words and many words form a speech signal. Even if the same words are spoken, the spectro-temporal dynamics of the signal will differ when the words are spoken by different speakers. This analysis of spectro-temporal patterns, in particular comodulation, helps to segregate a sound source from an acoustical background.

It is commonly assumed that processing of comodulation and interaural disparities is done by neural computation in the auditory system. Since, however, comodulation can also be used with only one ear, part of the beneficial effect of comodulation might also be explained by mechanical preprocessing at an early stage of the auditory pathway [Ernst and Verhey, 2006]. Before the auditory system can process the information in the acoustical stimulus, the physical pressure wave in the air has to be transformed into a neural signal. This task is performed by the non-neural part of the auditory pathway and a specialized type of sensory cells: the outer ear (pinna and ear canal), the middle ear (tympanic membrane and ossicles), and the inner ear consisting of the cochlea and the inner hair cells. The sound wave is acoustically filtered by the pinna and transmitted through the ear canal. The tympanic membrane transforms the pressure wave in the air into vibrations of the ossicles. The vibrations of the ossicles are translated into sound waves in the fluid-filled structure of the cochlea where the inner hair cells transform the mechanical signal into neural nerve pulses. Along this path, the cochlea plays the most important role. Cochlear processing not only includes the decomposition of the incoming signal into frequency components by a frequency-place transformation. Also active and nonlinear processes are present in the cochlea to improve the sensitivity to low intensity sounds and at the same time to enlarge the dynamic range of the system. Hence, the cochlea plays a fundamental role in the transformation of an external sound signal into an internal representation. All information that is lost at this stage cannot be used by successive processing stages. This means that an impairment of this preprocessing has an important impact on further processing and hence on the ability to perceive sound and to make use of signal cues.

Besides the important question to what extent the auditory system can use comodulation and interaural disparities separately, it is important to understand how a combination of these two cues is used by the system. In order to separate mechanical and neural processing, the contribution of cochlear processing to pschoacoustical effects is important. Depending on the transformation of the acoustical signal with its physical properties into an internal representation and the signal processing strategy, one cue in the signal might influence the performance of processing the other cue and vice versa. A systematic investigation of the single cues provides information about the implementation of the processing strategy to process a single cue and might give a indication of the internal representation of the signal. The systematic investigation of a combination of the cues can then provide information about the interrelation of the processing stages as implemented in the auditory system of mammals. A detailed model of the cochlea allows to estimate the contribution of cochlea preprocessing to psychoacoustically measured effects.

1.2 A review of the literature

In the following four sections, a brief review of the existing literature is given. A detailed review of the relevant literature is given in chapters 2, 3 and 4 in the context of the individual studies.

1.2.1 Processing of binaural cues

Early studies by Licklider [1948] on the intelligibility of a speech signal in the presence of noise showed that an interaural phase disparity of either the noise or the speech signal improved speech intelligibility. This experiment showed, that besides the ability to use temporal modulations in different frequency regions (i.e. at different positions along the basilar membrane) to detect a sound, it is an important capability of the auditory system to process differences between the signals at the two ears. Similar to CMR, binaural processing can be quantified by the ability to detect a masked signal. One of the first studies to investigate detection thresholds of tones in broadband noise was presented by Hirsh [1948]. The results of this study showed that detection thresholds are lower in conditions where the signal was interaurally shifted in phase (dichotic condition) compared to a condition where both signal and masker were interaurally in phase (diotic condition). The reduction in masking due to interaural disparities is often referred to as binaural masking level difference (BMLD) [Jeffress *et al.*, 1956]. Systematic investigation of the BMLD showed that the magnitude of the BMLD is dependent on signal frequency, masker bandwidth and interaural masker correlation [e.g. van de Par and Kohlrausch, 1999, van der Heijden and Trahiotis, 1997]. The BMLD is large for low frequencies and small masker bandwidths and tends to decrease for high frequencies and broadband maskers.
The conceptual models explaining the processing of interaural disparities can roughly be divided into two groups. One group is based on the approach of interaural cross correlation [e.g. Osman, 1971] and the other group on an equalization-cancellation concept proposed by Durlach [1963]. The cross-correlation model assumes that a decrease in interaural correlation when a dichotic signal is added to a diotic noise masker can be used as a cue to detect the signal. The equalization-cancellation (EC) model assumes that the auditory system is able to process

interaural disparities by two operations. The first operation is equalization of the disparities between the ears to interaurally match the masker or maximize the the interaural phase disparity of the signal. The second process is a cancellation of the two ear signals, cancelling signal components that are identical between the two ears, i.e. the masker, which leads to an improved internal representation of the signal. These models can be used to model data of simple masking experiments [e.g. Osman, 1971] up to speech intelligibility [Beutelmann and Brand, 2006].

Different physiological correlates of binaural processing have been proposed in the literature. A possible neural implementation of the cross-correlation approach are so-called delay lines [Jeffress, 1948]. A system of coincidence-detector neurons which are connected to axons with different lengths from the ipsi- and contralateral side and receive excitatory-excitatory projections (EE) allow the estimation of interaural delays. A neural implementation of the EC-model was proposed to be a system of cells which receive excitatory input from one ear and inhibitory input from the other ear (excitatory-inhibitory, EI). Such cells were found at the level of the inferior culliculus (IC) or at the level of the superior olivary complex (SOC) [McAlpine and Grothe, 2003, for a review].

An important aspect of the data which is also implemented in the models is the decrease of the BMLD towards higher frequencies. It is assumed that this decrease is the result of the reduced ability of the auditory system to code the finestructure of the signal [Zurek and Durlach, 1987]. This indicates that differences in interaural finestructure are used as a cue by the auditory system to improve the detectability of a masked signal. At higher frequencies, where finestructure cues cannot be used, a residual BMLD can be explained by processing interaural differences in the envelope of the signal.

1.2.2 Processing of across-frequency cues

An important property of the auditory system of many species is the ability to selectively process frequency components of an incoming sound. In mammals, this frequency selectivity is a consequence of the frequency-place transformation at the level of the cochlea in the inner ear. A powerful concept of processing in the auditory system is that the incoming sound is analyzed by a bank of overlapping bandpass filters, often referred to as critical bands or auditory filters [Fletcher, 1940]. A useful model for the description of masking experiments such as the detection of a tone in a masking noise is the power-spectrum model proposed by Fletcher [1940]. The main assumption of this model is that the signal-to-noise ratio of the signal and the masker in the auditory filter where the signal is located, determines the threshold. This means that masker energy outside the critical band does not contribute to the masking of the signal. This result suggests that the auditory system processes signals in independent channels. A limitation of this concept was demonstrated by Hall *et al.* [1984] with a masking experiment using a masker generated by a bandpass noise carrier modulated by a lowpass noise. The modulation of a bandpass noise with a lowpass noise introduces coherent level fluctuations, or comodulation, in the frequency regions of the masker. Hall *et al.* [1984] showned that thresholds of a tone in this modulated noise decreased when the bandwidth of the modulated noise exceeded the width of the auditory filter centered at the frequency of the tone, while thresholds of an unmodulated noise as masker remained constant for bandwidths larger than that of the auditory filter. This effect is often referred to as comodulation masking release (CMR). The data of this experiment cannot be explained by the power spectrum model. The conclusion of this experiment was that signal energy in remote filters can be used to improve the detection of a masked signal, i.e. that the auditory system makes use of across-channel processing. It was shown that frequency com-

ponents with a spectral distance of as much as three octaves or more can still be beneficial for signal detection in a comodulated noise masker [Cohen, 1991, Ernst and Verhey, 2006, 2008].
To investigate CMR, commonly two classes of experiment were used: the bandwidening experiment where the masker consists of a broadband noise centered at the signal frequency masker, and the flanking-band experiment where the masker consists of multiple narrow noise bands. It was argued that in the bandwidening type of experiment, part of the effect of comodulated maskers can be explained by so-called within-channel processing, i.e. by processing the contents of only one auditory filter [Schooneveldt and Moore, 1989, Verhey *et al.*, 1999, Buschermohle *et al.*, 2007, Ernst and Verhey, 2008]. The benefit which can be explained by within-channel processing in masking conditions with comodulated maskers is often referred to as within-channel CMR. In the flanking band paradigm, however, the results cannot solely be explained by within-channel processing [Verhey *et al.*, 1999, Piechowiak *et al.*, 2007] and the combination of the information from multiple auditory filters is necessary. This part of the CMR is often referred to as across-channel CMR. Dau *et al.* [2009] argue that CMR is not a consequence of an across-channel process since the benefit due to comodulated maskers can be abolished by embedding the remote frequency masker components into a temporal sequence of maskers. The CMR vanishes if the remote frequency components are outside a critical band, but not when all frequency components are within one critical band. One main difference between the study of Dau *et al.* [2009] and the previous studies is that the stimulus intervals in Dau *et al.* [2009] were composed of multiple temporally separated segments while in the other studies each stimulus interval consisted of one temporal segment only.
Possible physiological mechanisms underlying CMR have been proposed at different levels of the auditory pathway. It has been suggested that wideband inhibition at the level of the cochlea nucleus (CN) might be a physiological correlate of CMR [Neuert

et al., 2004]. An improved response to short tone pips in masker minima could be found if additional maskers were placed in inhibitory regions of a cell tuned to the signal frequency. Las *et al.* [2005] propose that CMR is the result of a suppression of phase-coherent responses of neurons at the level of the auditory cortex. In line with Las *et al.* [2005], the hypothesis that CMR is influenced by putting the masker components into a separate auditory stream [Dau *et al.*, 2009] suggest that CMR is processed at a high level of the auditory pathway. Next to these explanations, it has been hypothesized that at least part of the CMR can be accounted for at a low level of the auditory pathway. For certain stimulus conditions it has been proposed by Ernst and Verhey [2006, 2008] that CMR can partially be explained by suppression at the level of the cochlea. With the help of an effective model of the cochlea, Ernst and Verhey [2006] have shown that for certain stimulus conditions suppression of the masker at the signal frequency might effectively reduce the masking of the signal. Despite the diversity of the attempts to explain CMR, it is evident that the coherent level fluctuations in different frequency regions, i.e. the coherence of the temporal envelope in different frequency bands can be used as a cue by the auditory system to improve the detection of masked signals.

1.2.3 Processing of combined monaural and binaural cues

While processing of either binaural cues or the across-frequency cue of comodulation has been well investigated, there is only little data on the processing of a combination of interaural disparities and comodulation. Hall *et al.* [1988] measured detection thresholds of a 500 Hz pure tone which was masked by one narrowband noise masker alone and when an additional comodulated masker band was added with a center frequency of 400 Hz. The signal was presented either in phase between the two ears or out of phase. The second comodulated noise band provided information in the envelope in different frequency regions and the interaural disparity in the signal provided a cue in the finestructure for the binaural system. CMR was found to be larger in the diotic condition compared to the dichotic condition. Hall *et al.* [1988] concluded that comodulation cannot be used in dichotic listening conditions. Cohen [1991] measured detection thresholds for a 700 Hz tone in the absence and presence of an additional masker band centered at 600 Hz. The additional masker band was either comodulated or had uncorrelated envelope fluctuations and the signal was presented either in or out of phase between the ears. The results show that CMR was reduced in dichotic compared to diotic listening conditions, but a residual CMR was found when the uncorrelated masker condition was compared with the comodulated masker condition. This result indicates that a residual benefit due to comodulation can be found when the same spectra are used to evaluate CMR. In a more recent study by Hall *et al.* [2006], detection thresholds of an interaurally in- or out-of phase 500 Hz tone were measured in the absence or presence of several uncorrelated or comodulated noise bands with a spectral distance of 100 Hz. The results of this study show large individual differences and on average only a small CMR in dichotic conditions.

The results of these studies do not allow a simple conclusion about the ability of the auditory system to use comodulation in dichotic listening conditions. While data from Hall *et al.* [1988] indicate that no CMR can be found in dichotic listening conditions, data from Cohen [1991] and Hall *et al.* [2006] indicate that CMR and BMLD combine to some extent. The evaluation of CMR in dichotic listening conditions using different spectra might be difficult due to differences in the BMLD for maskers with different spectral contents. In addition, the evaluation of CMR with masker components both within and outside the critical band of the signal complicates the interpretation since the measured CMR can be a combination of both, within- and across-channel CMR. The above mentioned studies leave open the question if and how comodulation and interaural disparities can be used together by the auditory system in complex acoustical environments to form auditory objects. Together with the conclusions of the above mentioned studies, additional experiments with a systematic investigation of comodulation cues and interaural disparities can contribute to answering this question.

1.2.4 Mechanical preprocessing at the level of the cochlea

Cochlear processing has been subject to investigation for a long time. Early theories of Ohm [1843] and Helmholtz [1863] described the cochlea as a bank of independent resonators comparable to strings or oscillating beams. This more theoretical view of the cochlea was fundamentally revised by the work of v. Bekesy [1949b,a] who studied the oscillations of the basilar membrane in human cadaver cochleae. The conclusion of his studies was that there exists a traveling wave along the cochlea with a maximum response at different places along the membrane which depends on the input frequency. With this frequency dependent maximum at different places, the cochlea serves as a linear frequency analyzer. Later in-vivo studies of cochlea vibration in squirrel monkeys by Rhode [1971] showed that the peak of the travelling wave is much sharper than observed by v. Bekesy [1949b,a] and that the shape of the travelling wave changes with stimulus amplitude, i.e. is nonlinear. It was later shown by [Kemp, 1979] that the cochlea emits sounds with and without external stimulation, so-called otoacoustic emissions (OAEs). As an explanation of the sharp peak for low stimulus levels and for the OAEs, the idea that an active process might be involved in cochlear processing proposed by Gold [1948] was reviewed, and Davis [1983] termed the active process in the cochlea the "cochlea amplifier". This assumption of

an active process was supported by deBoer [1983] who showed mathematically that an energy source is a necessary condition to obtain a sharp peak as observed in the data by Rhode [1971]. The first stable model incorporating such an active process by a negative damping term was proposed by Neely and Kim [1983] who implemented negative damping to mimic a feed-forward mechanism along the basilar membrane. A more quantitative analysis of the active process was made possible by the inverse solution of the cochlea vibration patterns. Zweig [1991] and deBoer [1995] extracted the mechanical parameters of the cochlea from measurements of the basilar membrane motion. Since then cochlea models have been refined on the basis of data from OAE experiments. Zweig and Shera [1995] and Talmadge *et al.* [1998] showed that so-called finestructure effects as for example found in various kinds of OAE measurements can also be modelled with a nonlinear and active model of the cochlea. More recent studies introduced the concept of Hopf bifurcations into the description of the cochlea amplifier [Hudspeth, 2009, for a review]. For the analysis of the nonlinearity, concepts from system theory were also applied to analyze the behavior of nonlinear cochlea models [Elliott *et al.*, 2007].
The increase in computational power during the last years encourage the extension of the modeling efforts of the last decades. The filterbank models of the auditory periphery commonly used in auditory models might be replaced by more realistic models of the cochlea to investigate the influence of cochlear preprocessing on auditory perception and object formation.

1.3 Aim and scope of this thesis

The aim of this thesis is to contribute to the understanding of the mechanical and neural processing strategies within the auditory system to use combined across-frequency and binaural cues and their role in auditory object perception in humans.

There are different ways of investigating the auditory system. One way is the investigation of the system on a physiological and anatomical basis, either at a cellular level or with imaging techniques. Another approach are behavioral experiments performed in psychoacoustics where sounds with well defined properties are played to a listener and the effect of these sounds is indirectly evaluated through the response of the listener. A third approach, which is dependent on data from the other two fields, is the domain of models. In this thesis, psychoacoustical experiments with humans were performed, accompanied by conceptual and more realistic modeling approaches to test the hypothesized processing strategies which were drawn from the experimental results.

This thesis is divided into three separate studies. In chapter 2, a psychoacoustical experiment combining the effects of CMR and BMLD in combination with a simplified model of the auditory system was developed and used to evaluate the ability of the auditory system to combine binaural finestructure cues and monaural across-frequency amplitude cues. The results of the first part were generalized for a broad frequency range in chapter 3. In chapter 4, the nonlinearity of a physical cochlea model was developed to account for the effects of stimulus level and multi-tone stimulation as found in physiological and psychophysical data. This model was then applied to a modulation detection paradigm in order to investigate the consequences of nonlinear peripheral processing on the internal representation of amplitude modulated tones. With the model in connection with a simple decision stage as an artificial observer, psychoacoustical data on modulation detection could be simulated.

2 Superposition of masking releases[1]

Abstract

We are constantly exposed to a mixture of sounds of which only few are important to consider. In order to improve detectability and to segregate important sounds from less important sounds, the auditory system uses different aspects of natural sound sources. Among these are (a) its specific location and (b) synchronous envelope fluctuations in different frequency regions. Such a comodulation of different frequency bands facilitates the detection of tones in noise, a phenomenon known as comodulation masking release (CMR). Physiological as well as psychoacoustical studies usually investigate only one of these strategies to segregate sounds. Here we present psychoacoustical data on CMR for various virtual locations of the signal by varying its interaural phase difference (IPD). The results indicate that the masking release in conditions with binaural (interaural phase differences) and across-frequency (synchronous envelope fluctuations, i.e. comodulation) cues present is equal to the sum of the masking releases for each of the cues separately. Data

---

[1]With kind permission from Springer Science+Business Media: Epp, B. and Verhey, J.L. **(2009a)** "Superposition of masking releases." J. Comput. Neurosci.(26), 393-407 (2009). © 2009, Springer.

and model predictions with a simplified model of the auditory system indicate an independent and serial processing of binaural cues and monaural across-frequency cues, maximizing the benefits from the envelope comparison across frequency and the comparison of fine structure across ears.

2.1 Introduction

A fundamental property of the auditory system of many species is its frequency selectivity which is a consequence of the frequency place transformation on the basilar membrane. In psychoacoustics this frequency selectivity is usually characterized using detection data of sinusoidal signals masked by Gaussian noise. Many aspects of the masking data can be understood by assuming that the auditory system analyzes the incoming sound by a bank of overlapping bandpass filters and that the ratio of the intensity of the signal plus masker and the signal alone in the filter centred at the signal frequency determines threshold. However, in more natural environments, the auditory system seems to use more elaborate strategies for the detection of sounds. Such strategies are thought to be evolutionary developed as an important survival strategy [Nelken *et al.*, 1999]. One strategy is the analysis of disparities in the fine structure of the signals arriving at the two ears, as e.g. used to localize sound sources. In psychoacoustics the benefit of the auditory system due to this binaural information is often described as a binaural masking level difference [BMLD; e.g. Jeffress *et al.*, 1956]. The BMLD is defined as the difference in threshold of a condition without and a condition with disparities in either the signal or the noise between the ears.
Apart from the processing across ears, the auditory system seems to use the similarity of level fluctuations in different frequency regions within one ear as an additional detection cue. One effect related to this ability is comodulation masking release [CMR; Hall *et al.*, 1984] where it is found that signal detection improves if the masker has coherent level fluctuations across frequency, i.e. is comodulated. A common reference to quantify CMR is a masking condition with the same masker spectrum as in the comodulated condition but with incoherent level fluctuations in different frequency regions (uncorrelated condition).

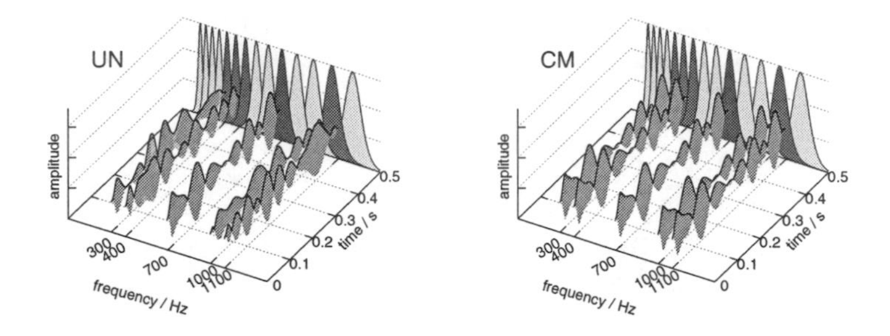

Figure 2.1: Schematic spectrogram of the two conditions with different masker envelope correlation across frequency that were used in the present study. Masker envelopes are uncorrelated in the left panel (UN condition) and comodulated in the right panel (CM condition). On the back plane of each panel, a gammatone filter bank is shown. Shades of grey represent the amount of masker energy within the respective filter, dark grey referring to a high excitation. The masker in the middle is the masker component centred at the signal frequency (Signal centred band, SCB).

An example for the two conditions is shown in Figure 2.1. Both, CMR and BMLD can amount to 10 dB or more depending on the properties of the signal and the masker (e.g. van de Par and Kohlrausch [1999] for BMLD, Verhey *et al.* [2003] for a review on CMR). In psychoacoustics similar processing strategies have been proposed for CMR and BMLD [Buus, 1985, Richards, 1987, Green, 1992, van de Par and Kohlrausch, 1999], even though the underlying time scales are different. In general, BMLD is thought to be due to fine structure disparities in the signals, which makes it necessary to analyze signals with a time resolution in the millisecond range. In contrast, the across-frequency comparison underlying CMR is sensitive to the level fluctuations represented by the temporal envelope of the signals at different

locations along the basilar membrane, demanding a much lower time resolution.
Several physiological studies investigated the neural mechanism underlying information processing of binaural cues [see e.g. Kapfer *et al.*, 2002, Thompson *et al.*, 2006, Darrow *et al.*, 2006] and across frequency cues [CMR; see e.g. Nelken *et al.*, 1999, Nieder and Klump, 2001, Pressnitzer *et al.*, 2001, Neuert *et al.*, 2004, Las *et al.*, 2005, Li *et al.*, 2006]. Whereas there seems to be general agreement about the nuclei involved in binaural processing (i.e. the stages where differences in the signals of two ears are processed), it is still unclear which stages of the auditory pathway are involved in across-frequency processing. Some studies regard broadband inhibition at the level of the cochlear nucleus as the neural mechanism underlying CMR [Pressnitzer *et al.*, 2001, Meddis *et al.*, 2002, Neuert *et al.*, 2004]. Others argue that the neural correlate of CMR is envelope locking suppression [Nelken *et al.*, 1999]. Las *et al.* [2005] suggest that a CMR effect might be present at the level of the IC and that the tone is represented much more explicitly at the level of the auditory cortex by the suppression of envelope locking.

It was also hypothesized, that part of CMR may be explained by nonlinear peripheral processing. Buschermohle *et al.* [2007] presented a model based on data of neural recordings from starlings. This model assumes a level-independent compression and the covariation of the mean firing rate of neural populations with the mean compressed envelope. The model predicts parts of CMR measured in different experiments by evaluation of changes in the mean compressed envelope. Apart from compression, also suppression at a cochlea level has been proposed to account for a part of CMR [Ernst and Verhey, 2006, 2008]. For large level differences between on- and off-frequency masker components, most of the CMR could be accounted for by a model incorporating a suppression stage.

The aim of the present study is to shed some light on the topographic organization and connections between the nuclei responsible for the analysis of coherent level fluctuations across frequency and binaural information using psychoacoustical experiments where both cues are provided.

There are only a few studies investigating the combined effect of comodulation and binaural cues [Hall *et al.*, 1988, Cohen, 1991, Hall *et al.*, 2006]. Those studies obtained contradictory results.

Hall *et al.* [1988] found no interaction when both cues were present. Their data indicated that the cue providing the largest masking release determined thresholds.

Their results indicate that comodulation is analyzed in a different pathway than interaural differences (i.e. they are parallel processes). However, the data of more recent studies are not in agreement with the hypothesis of a parallel processing. Cohen [1991] found that a small masking release due to comodulation can be produced even when the detectability of a stimulus has already been improved due to binaural cues. In a more recent study, experimental data was compared to predictions using a signal-detection-theory approach [Hall *et al.*, 2006]. Under the assumption of independence and optimal combination of CMR cues and BMLD cues, Hall *et al.* [2006] found that the benefit in situations with both cues present is larger than predicted by an integration model [Green and Swets, 1988].

One problem of the studies is that they used a small frequency separation between on- and off-frequency masker components. Thus, it is unclear if their CMR is solely due to true across-frequency processes. In addition, two of the studies [Hall *et al.*, 1988, 2006] only calculated the CMR as a difference between conditions with different spectra which may obliterate the results since it is known that the BMLD depends on the masker bandwidth [van de Par and Kohlrausch, 1999].

In the present study we hypothesize that the combined masking release is close to the sum of the masking release produced by the cues alone, provided that the monaural masking release is mainly due to across-frequency processes and the appropriate reference is used. We further hypothesize that the data is in line with a serial processing of the across-frequency cue of comodulation and the binaural cue of interaural phase difference (IPD). Provided that the first stage does not alter the information needed in the second stage, such a serial alignment of the processing stages should result in an addition of the effects.

In order to test these hypotheses, data and model predictions of experiments combining cues assigned to CMR and BMLD are presented. In contrast to previous studies, the two cues were varied systematically by using various interaural phase differences of the signal and by using different portions of comodulated masker components. A frequency separation between on- and off-frequency masker components wider than in the previous studies was used to avoid a large contribution of within-channel cues.

A minimum frequency separation between on- and off-frequency masker components of more than half an octave was used to avoid within-channel cues in CMR. At such a spectral distance, a model using compresssion predicts CMR only if an unrealistic broad peripheral filter is assumed [Buschermohle *et al.*, 2007] and a model using modulation cues also predicts negligible CMR [Piechowiak *et al.*, 2007]. A substantial effect of suppression on CMR was found for low-level signal-centred bands and high-level flanking bands. In order to reduce a possible influence of suppression the same level was used for the signal-centred band and the flanking bands.

In previous CMR flanking band experiments essentially two different types of masking noise were used which differ in their envelope statistics: multiplied noise and Gaussian noise. It was shown that thresholds of a sinusoidal signal depend on the envelope amplitude statistics of the masker and may also result in different magnitudes of the masking release [Moore *et al.*, 1990, Ernst and Verhey, 2006, Verhey *et al.*, 2007].

Figure 2.2 shows the envelope distributions and time waveforms of samples of these two noise types. Multiplied noise is generated by multiplying a lowpass noise with a sinusoid at the desired center frequency. This noise type has an envelope distribution corresponding to the positive half of a Gaussian distribution. Multiplied noise was often used in early CMR experiments, presumably because it allowed an easy analogue implementation of bandpass noises with a steep roll off at the high and low-frequency sides. The second masker type is Gaussian noise with an envelope statistic following a Rayleigh distribution. This masker type was commonly used in e.g. spectral masking and binaural experiments in psychoacoustics and for broadband excitation of neurons in neurophysiology. Fluctuations of this masker type result from beatings between the frequency components rather than from an imposed modulation resulting in a lower modulation depth than for the multiplied noise.
In order to facilitate the comparison with CMR data from literature, both masker types are used in the present study. In addition, the use of both masker types might show a possible influence of the masker statistics on the combination of CMR and BMLD. The experimental part of the present study is divided into two parts. In a first experiment, the binaural cue was systematically varied with (comodulated condition) or without (uncorrelated condition) a monaural across-frequency cue. In a second experiment, the monaural cue was systematically varied for three different interaural differences of the signal. On the basis of the experimental results, a computational model of the auditory system was developed and matched to the maximum CMR for a diotic signal and masker and the BMLD for an

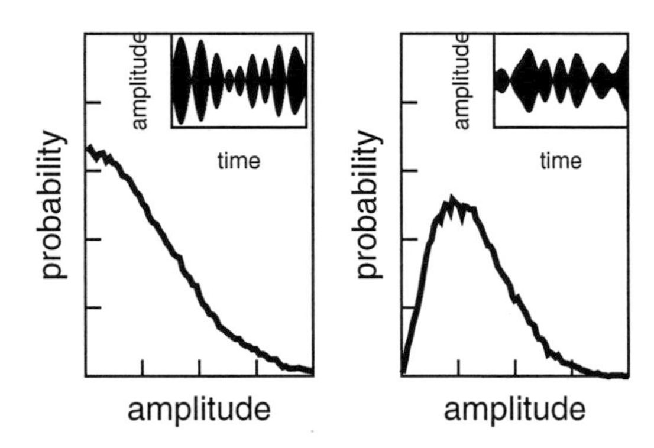

Figure 2.2: Envelope amplitude statistics of the noise masker types used in the experiment. The left panel shows a sample interval of a multiplied noise masker and the right panel of a Gaussian noise masker. The multiplied noise masker envelope amplitude statistics corresponds to the positive half of a Gaussian distribution while the Gaussian envelope amplitude statistics follows a Rayleight distribution.

antiphasic signal in the presence of a diotic noise masker with uncorrelated envelopes independently. The conceptual model was used to quantitatively predict experimentally obtained intermediate data points.

The experimental results show a clear combination of the two effects of CMR and BMLD, supporting the hypothesis of serially aligned processing stages. This hypothesis is further supported by the conceptual model that assumes such a serial alignment of the stages responsible of across-frequency and binaural processing using a minimum set of parameters.

2.2 Methods

2.2.1 Procedure

A three interval, three-alternative forced-choice procedure with adaptive signal-level adjustment was used to determine the masked threshold of the sinusoidal signal. The three intervals in a trial were separated by gaps of 500 ms. Subjects had to indicate which of the intervals contained the signal. Visual feedback was provided after each response. The signal level was adjusted according to a two-down, one-up rule to estimate the 70.7% point of the psychometric function [Levitt, 1971]. The initial step size was 8 dB. After every second reversal the step size was halved until a step size of 1 dB was reached. The run was then continued for another six reversals. From the level at these last six reversals, the mean was calculated and used as an estimate of the threshold. The final threshold estimate was taken as the mean over four threshold estimates.

2.2.2 Stimulus and apparatus

To investigate the combination of CMR and BMLD it is advantageous to have large masking releases and masking releases of similar size. This facilitates the interpretation of the combination of the effects. In order to obtain a substantial BMLD, the signal frequency should be below about 1.5 kHz [van de Par and Kohlrausch, 1997]. Another constraint to be met is a certain spectral distance between the channels containing the signal and the off-frequency masker components in order to reduce within-channel CMR. Thus a signal frequency of 700 Hz was chosen which is the same as used by Hall *et al.* [1990].

The signal was temporally centred in the masker and had a duration of 250 ms including 50-ms $cos2$ ramps at signal on- and offset. The masker duration was 500 ms including 50-ms $cos2$ ramps at masker on- and offset. [2] The masker consisted of five noise bands. Each noise band had a bandwidth of 24 Hz and a level of 60 dB SPL. One noise band was centred around the signal frequency, and the other four noise bands were centred at

[2] An asynchronous onset and offset of the signal and masker is commonly referred to as a fringe condition. Hatch *et al.* [1995] compared a synchronous condition with a condition where the signal was gated while the masker was presented continuously. The latter continuous-masker condition may be regarded as a condition with an infinite fringe. They reported a gating effect of up to 6 dB on the amount of CMR depending on the number and the bandwidth of the masker bands. For a condition comparable to the one used in the present study, they measured a gating effect of 3 dB on average for five bands and a bandwidth of 20 Hz. However, they derived the CMR comparing the threshold for the comodulated condition with a condition where only the signal-centred band was present. McFadden and Wright [1992] showed that the effect of an asynchronous onset for signal and masker is considerably larger (by several dB) when the CMR is calculated relative to the signal-centred-band only condition than when the CMR is determined calculating the difference between uncorrelated and the comodulated condition as in the present study. Thus, it is unlikely that the asynchronous gating of signal and masker had an influence on the magnitude of CMR in the present study.

300 Hz, 400 Hz, 1000 Hz and 1100 Hz. Assuming a fourth order gammatone filter as the auditory filter centred at 700 Hz, the closed of these noise bands are attenuated by 40 dB, i.e. it is likely that most of the CMR is due to across-channel processes. For this set of flanking bands, Hall *et al.* [1990] obtained a reasonable sized monaural CMR of about 12 dB. This magnitude is compareable to the average magnitude of the BMLDs for antiphasic signals with frequencies in the range from 500 Hz to 1 kHz and diotic broadband noise maskers [van de Par and Kohlrausch, 1999]. The noise bands had either uncorrelated intensity fluctuations (UN condition, left panel of Figure 2.1) or coherent intensity fluctuations (comodulated, CM condition, right panel of Figure 2.1). The masker was always presented diotically. The signal had an interaural phase difference (IPD) in the range from 0 to 180 degrees. In the first experiment, the IPD was either 0, 14.4, 36, 72, 144, or 180 degrees[3]. Two noise types were used: multiplied noise and Gaussian noise (see Figure 2.2).

Multiplied noise masker components were generated by multiplying a random phase sinusoidal carrier at the desired center frequency with a narrowband noise without a DC component.

For the uncorrelated (UN) condition, independent realizations of the noise modulator were used, while in the comodulated (CM) condition, the same modulator was used for all masker components.

Gaussian noise masker components were generated in the frequency domain by assigning numbers gained from draws from a normally distributed process to the real and complex parts of the desired frequency components in the frequency bands. This

---

[3]The IPD of 144 degrees originates from a geometrical approximation of the travel distance of the sound wave, resulting in a number of multiples of $\frac{1}{25}\pi$. The intermediate IPD between 0 and 144 degrees were generated by halving this IPD.

was done independently for the uncorrelated (UN) condition, while the numbers of one single draw are assigned to all frequency bands in the comodulated condition. The real part of the subsequent inverse fast Fourier transform yielded the desired waveform with uncorrelated and comodulated across frequency envelope structure, respectively.

For the second experiment, only multiplied-noise maskers and three IPDs of the signal were used: 0, 72, and 180 degrees. In contrast to the first experiment, thresholds were measured for various degrees of comodulation using different portions of comodulated and uncorrelated noise maskers. They were generated by a weighted sum of uncorrelated and comodulated maskers:

$$\text{Masker}_{\alpha} = (1-\alpha) \cdot \text{Masker}_{UN} + \alpha \cdot \text{Masker}_{CM} \qquad (2.1)$$

Thus, the portion of comodulation was controlled by the parameter α, ranging from 0 to 100% in steps of 25%. For example, the masker for $\alpha = 0$ corresponds to the uncorrelated masker, while for $\alpha = 1$ the masker corresponds to the comodulated condition. Under certain assumptions, the value α can be related to the cross covariance between the envelopes of the on-frequency masker (o) and a flanking band (f) in an analytical expression. Assuming infinitely long signals, statistical independence of the envelopes of the masker bands, identical mean and energies, the cross covariance can be expressed as:

$$r_{o,f} = \frac{\alpha}{\sqrt{(\alpha^2 + (1-\alpha)^2)}} \qquad (2.2)$$

All signals were generated digitally with a sampling frequency of 44100 Hz using MATLAB. Signals were A/D converted (RME ADI-8 DS), amplified (Tucker Davis Technology HB7) and presented to the test subjects in a double-walled sound attenuating booth via Sennheiser HDA200 headphones.

2.2.3 Subjects

Eight listeners with normal hearing participated in the experiment, varying in age from 24 to 29 years, five females and three males. None of the listeners had any history of hearing difficulties and their audiometric thresholds were 15 dB HL or less at the audiometric frequencies 125, 250, 500, 1000, 2000, 4000, and 8000 Hz. The subjects had at least 4 h experience in CMR experiments before collecting the data.

2.2.4 Model

The model is structured in three stages according to Figure 2.3. Implementation is motivated by a hypothetical neural circuit as shown in the left panel of Figure 2.3. A preprocessing stage is followed by serially aligned across-frequency and across-ear stages. After preprocessing by the basilar membrane, the across-frequency processing is implemented as wideband inhibitor that receives excitatory input from a large frequency range.The wideband inhibitor provides inhibitory input to a narrow-band cell tuned to the frequency of the target (BF), i.e. the cell which is most excited by the presence of the signal. Across-ear processing is shown as an excitation-inhibition (EI) cell, being excited by the ipsilateral ear and inhibited by the contralateral ear.

The combined effect of comodulation and interaural disparities on signal detection is implemented as common excitation of a summation neuron (S) by the two across-frequency stages (ipsi- and contralateral) and the output of the EI cell.

The right panel of Figure 2.3 shows the functional realisation of the model as used in the present study. The model is economical, in the sense that it is specified by essentially two parameters: the form of the weighting function (w(t)) and the performance of the binaural stage (b). The outputs of the different stages of the model for the masker only (blind interval, black) and the masker plus super-threshold signal (target interval, grey) are shown in Figure 2.4 for the uncorrelated (left panels) and correlated condition (right panels), respectively.

The preprocessing within the model is done for the left and right channel separately and consists of three parts. At first the signal is filtered by a bank of overlapping gammatone filters [Hohmann, 2002]:

$$Y = \Re\{x * H\} \tag{2.3}$$

where x is the vector containing the original signal. The convolution with the matrix H with the impulse responses h_k yields the matrix Y of signals within the pass-bands of the single filters tuned to frequency k. The elements of the matrix Y are given by:

$$y_{uk} = \sum_{j=-\infty}^{\infty} x[j] \cdot h_k[u-j] \tag{2.4}$$

The output of each filter is half-wave rectified and lowpass-filtered with a cut-off frequency of 770 Hz. These steps represent the first order approximations of the frequency selectivity on the basilar membrane and the mechano-electric transduction at the level of the haircells:

$$Y_R = \frac{Y + |Y|}{2} \tag{2.5}$$

The half-wave rectified signals in Y_R are filtered using a fifth order lowpass-filter with a cut-off frequency of 770 Hz (as for example used in Breebaart *et al.* [2001]):

$$Y_{R_{LP}} = Y_R * h_{770} \tag{2.6}$$

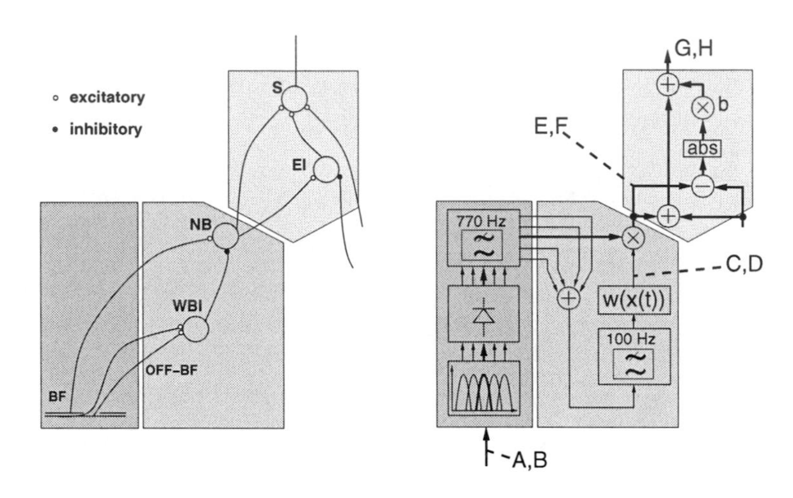

Figure 2.3: Schematic structure of the model. In the left picture, a hypothetical neural structure is shown. The right picture shows the functional implementation of the model. The first stage (bottom left) consists of a preprocessing stage, implemented as a gammatone filter-bank, a half-wave rectification and a lowpass filtering. The following across-frequency stage (bottom right) uses across-frequency information to modify the internal representation in the filter around the target frequency. This is implemented by computation of a weighting function which is applied to the output of the filter centred at the target signal frequency. The first two stages are applied to the acoustical signal of each ear separately. In the binaural stage (upper box), a difference detector is situated. In the functional model the output of the two ears is subtracted, amplified and added to the sum of the two ear signals. The capital letters on the functional scheme refer to the panels of the next figure showing the signal at different stages of the model.

where h_{770} represents the vector containing the filter coefficients of the lowpass filter. The result given in $Y_{R_{LP}}$

$$y_{R_{LP_{uk}}} = \sum_{j=-\infty}^{\infty} y_{R_{uk}}[j] \cdot h_{770}[u-j] \quad (2.7)$$

contains a half-wave rectified and lowpass filtered internal representation of the single passbands of the filterbank.

Across-frequency interaction is incorporated within the model by a summation of the output of the off-frequency filters, i.e. excluding the filter that is tuned to the signal frequency:

$$y_{adj} = \sum_{k=1}^{t-1} Y_{R_{LP_k}} + \sum_{k=t+1}^{N} Y_{R_{LP_k}} = \sum_{k=1,k\neq t}^{N} Y_{R_{LP_k}} \quad (2.8)$$

where $Y_{R_{LP_k}}$ is the signal in the k$-th$ filter and t the index of the filter tuned to the signal frequency. This results in the sum signal y_{adj} containing across-frequency information. This stage is a simplified version of a wideband (WBI) cell shown in Figure 4. The frequency response of this stage is flat except for a decrease in sensitivity by a maximum of 6 dB at the signal frequency. This summed output is low-pass filtered with a cutoff frequency of 100 Hz in order to extract the average slow envelope fluctuations in adjacent bands:

$$y_{adj_{LP}} = y_{adj} * h_{100} \quad (2.9)$$

with the impulse response of the lowpass filter given by h_{100}. The 100-Hz lowpass filter can be regarded as a modulation-lowpass filter, which was hypothesized in effective models of modulation perception [Viemeister, 1979, Strickland and Viemeister, 1996, Ewert and Dau, 2000]. The resulting across-frequency envelope $y_{adj_{LP}}$, representing the average envelope fluctuations in adjacent bands, is used to compute the weighting function w in the form:

$$w = 10^{-a \cdot y_{adj_{LP}}} \; ; \; a = 60 \quad (2.10)$$

For the uncorrelated condition, the weighting function is nearly constant i.e. it attenuates the masker and signal in the filter centred at the signal frequency by a similar amount (Figure 2.4 C). For the comodulated condition, the weighting results in an attenuation that is more efficient for the masker than for the signal since the weighting is low whenever the masker intensity is high. In contrast, the signal in the minima of the masker envelope will be enhanced resulting in a higher signal-to-noise ratio in the comodulated condition (Figure 2.4 D). The weighting function is applied to the output of the filter whose best frequency equals the target frequency. This across-frequency stage can be regarded as an effective realisation of wideband inhibition at the level of the cochlear nucleus which was hypothesized as a neural mechanism underlying CMR by [Pressnitzer *et al.*, 2001, Neuert *et al.*, 2004].

The output of the across-frequency stage y_{AF} is then computed by:

$$y_{AF_u} = Y_{R_{LP_{ut}}} \cdot w_u \tag{2.11}$$

resulting in a weighting of the instances in the target filter with the weighting function. This preprocessing applied to the left and right ear signal results in the two signals y_{AF_L} and y_{AF_R} representing the output after across-frequency processing on the left and right ear, respectively. The resulting output is shown in Figure 2.4 E (uncorrelated) and F (comodulated).

The acoustic signals of the right and left ear are processed separately using the previously described monaural processing stages. In the binaural stage, the two monaural pathways are combined into a single pathway. The binaural stage detects absolute interaural differences D by subtracting left and right channel and calculating the absolute value:

$$D = |y_{AF_L} - y_{AF_R}| \tag{2.12}$$

This is a simplified implementation of an a cancellation process within an equalisation-cancellation model as proposed in

Durlach [1963]. For the current study, an equalisation process preceding the cancellation is not necessary since only diotic maskers are used. This difference is amplified and added to the summed output of the CMR stages such that the final internal representation Y_{target} is given by summation of the monaural channels and the binaural gain:

$$y_{target} = (y_{AF_L} + y_{AF_R}) + b \cdot D \tag{2.13}$$

with the parameter $b = 1.7$ regulating the efficiency of the binaural stage. In the case of no interaural differences, the difference D will be zero, whereas it will be maximal for an IPD of the target signal of 180 degrees (see Figure 2.4 G, H).

The experiment was simulated with the same procedure as used for the subjects to estimate thresholds. In the simulated runs the model served as an artificial observer. As a decision variable, the steady-state response of the model output was integrated and corrupted by a normally distributed multiplicative noise $\mathcal{N}(1, \sigma^2)$ with a standard deviation of $\sigma = 2\%$. The interval with the largest magnitude of the decision variable was chosen by the artificial observer[4].

The adaptive procedure was simulated 100 times. The mean and standard deviation was calculated from these 100 runs and used as an estimate of the predicted threshold and it's accuracy.

The model parameters a, b and σ were chosen to simulate mean data over all subjects for the CMR for zero IPD and for the

[4]Another approach commonly used in psychoacoustic models is to calculate the signal-plus-noise-to-noise ratio, i.e. the ratio of the model output for the combination of signal and noise and for the noise alone. If this ratio exceeds a certain critical value, the signal is assumed to be detected [Fletcher, 1940, Ewert and Dau, 2000, Oxenham, 2001, Plack *et al.*, 2002]. To adapt such an approach to a 3AFC procedure as used here, one can use model output of the signal interval compared to the maximum of the output to the noise intervals [Ernst and Verhey, 2006].

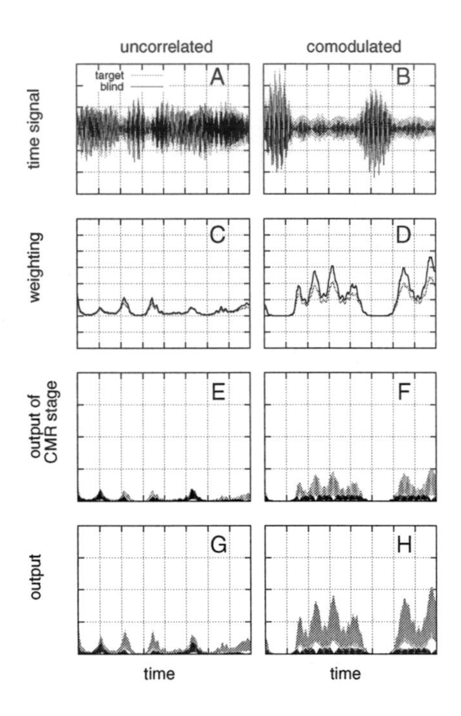

Figure 2.4: Internal representations within the model. Panels in each row show the internal representation at different stages of the model. The left column shows internal representations for the uncorrelated (UN) condition and the right column shows the representations for the comodulated (CM) condition. All panels show signals for a masker only interval (blind interval, dark grey) and for a target interval (light grey), where a signal was added to the masker (5dB above the level of the noise band around the signal frequency). The first row (A,B) shows acoustical signals. The second row (C,D) shows the resulting weighting function. In the third row (E,F) the output of the CMR stage is shown. The last row (G,H) shows the final contents of the target filter.

BMLD for an antiphasic signal in the presence of an uncorrelated multiplied noise masker. In Figure 2.5, experimental results (left panels) for these conditions are shown together with the corresponding simulated thresholds of the adjusted model (right panels). The upper panels show thresholds, the middle panels the difference between uncorrelated and comodulated thresholds (CMR) and the lower panels the binaural gain (BMLD). Modelled thresholds are in good agreement with experimentally obtained data. The same parameters were used for the predictions of all other data points.

2.3 Results

2.3.1 CMR as a function of the IPD

The left column of Figure 2.6 shows mean results for the multiplied noise masker of the eight subjects who participated in the experiment. In the top panel, detection thresholds for the uncorrelated (triangles) and comodulated (squares) condition are shown. Thresholds are expressed in dB relative to the level of the masker band centred at the signal frequency (signal-centred band, SCB). The middle panel shows the mean values and the standard deviations for the CMR. The mean and standard deviation of the CMR was derived on the basis of the individual CMR. In the bottom panel, the benefit due to the interaural phase shift of the signal is shown as a binaural masking level difference. As for the CMR, mean and standard deviations are based on the individual binaural masking level differences.

The threshold for the uncorrelated condition with no IPD is 2 dB. Thresholds decrease monotonically as the IPD increases. The lowest threshold for the uncorrelated condition is -10 dB for an IPD of 180 degrees. The comodulated threshold for an IPD of 0 degree is -10 dB. As for the uncorrelated condition, thresholds for the comodulated condition decrease as the IPD increases.

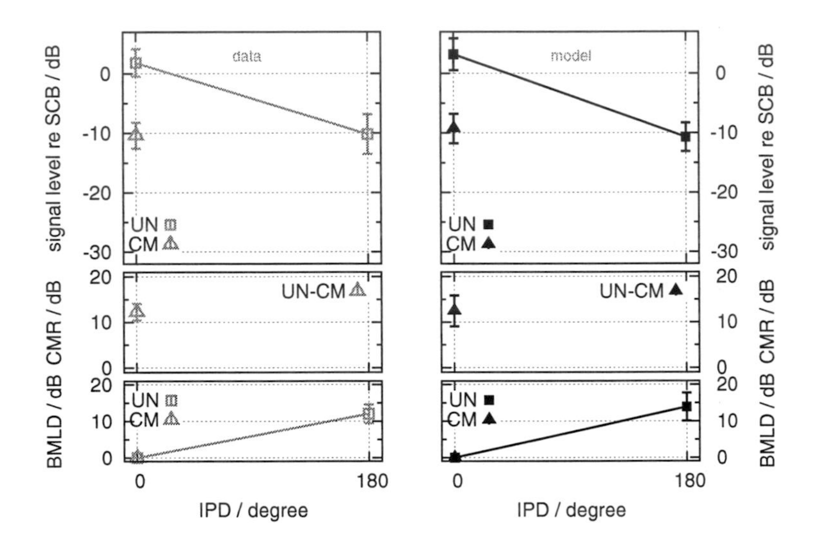

Figure 2.5: Model predictions (right panels) adjusted to experimentally obtained data (left panels). The upper row shows thresholds for uncorrelated (UN) maskers (squares) and comodulated (CM) maskers (triangles). In the middle panels, CMR is plotted as the difference of uncorrelated and comodulated thresholds. The lower panel shows the difference between diotic (no interaural phase difference, IPD) and dichotic thresholds as the BMLD. All points are plotted as a function of the IPD. The model is adjusted for a diotic signal to CMR and for an uncorrelated multiplied noise masker with an antiphasic signal to BMLD.

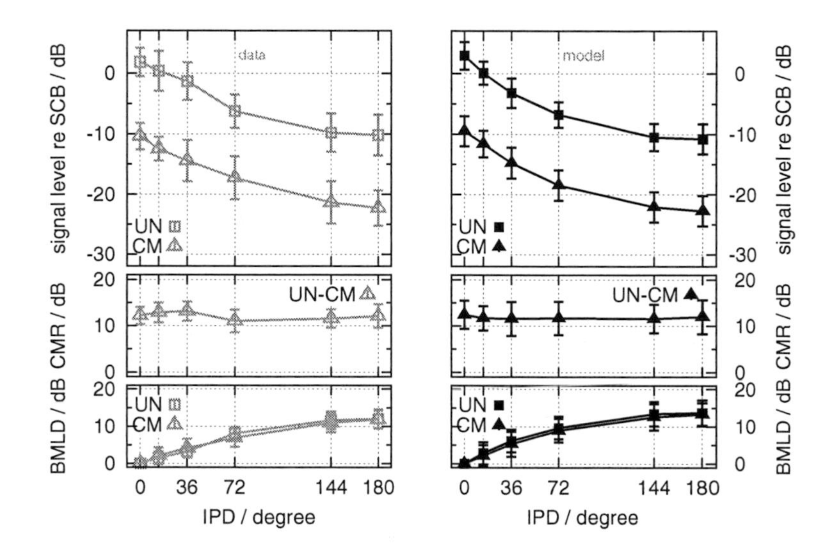

Figure 2.6: Measured thresholds (grey symbols) together with simulated thresholds (black symbols). Results for multiplied noise maskers in the conditions UN (squares) and CM (triangles) are shown. In the upper row thresholds are plotted relative to the level of the SCB. In the middle panel the resulting CMR is shown. The lower panel shows the BMLD for conditions UN (squares) and CM (triangles). All curves are plotted as a function of the interaural signal phase difference (IPD).

The lowest threshold for the comodulated condition is -22 dB for an IPD of 180 degrees. The difference between thresholds for the uncorrelated and comodulated condition at each value of the IPD hardly changes with IPD. This is reflected in the constant CMR of about 12 dB shown in the middle panel of Figure 2.6. Consequently, the increase in the BMLD as the IPD increases is similar for the uncorrelated and comodulated condition (lower panel of Figure 2.6). For both conditions, the maximum BMLD is about 12 dB for an IPD of 180 degree. In general, the standard deviations for the CMR and the BMLD are smaller than the standard deviations of the thresholds, indicating that sujects have different sensitivities but still show similar masking releases.

The right column of Figure 2.6 shows simulated thresholds. The same symbols were used for model predictions and experimental data (shown on the left panels of Figure 2.6). In general, model predictions agree with experimental data. In particular, the model is able to quantitatively predict the decrease of thresholds with increasing IPD. It also predicts the same CMR for all IPDs, in agreement with the data. Apart from the matched points shown in Figure 2.5, the course of the thresholds as a function of the IPD as well as the constant CMR are true predictions of the model.

Figure 2.7 shows experimental data (left panels) and model predictions (right panels) for the Gaussian noise masker. As in Figure 2.6 the top panels show thresholds, the middle panels show the CMR and the bottom panels show the BMLD. The average measured threshold for the uncorrelated condition with zero IPD of 4 dB is slightly higher than for the multiplied noise masker (2 dB). Thresholds decrease as the IPD increases with a minimum threshold for the uncorrelated condition at an IPD of 180 degree of -9 dB. The maximum BMLD is 13 dB, i.e. essentially the same maximum BMLD is measured for the two masker types that were used in the present study. The threshold

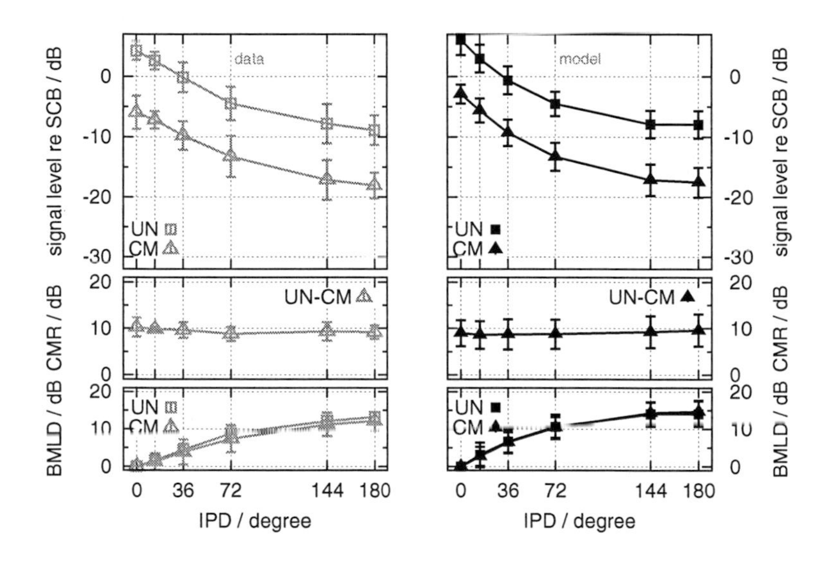

Figure 2.7: Same as Figure 2.6 for Gaussian noise maskers.

for the diotic signal masked by a comodulated noise is -6 dB. As for the other noise type, thresholds for the uncorrelated and comodulated conditions decrease as the IPDs increase. The lowest threshold for the comodulated condition is -18 dB at an IPD of 180 degree. The CMR is shown in the middle panel of Figure 2.7. It amounts to 10 dB at an IPD of 0 and is almost constant for all values of the IPD. Thus the CMR is 2 dB smaller than for the multiplied noise. For both conditions the BMLD increases as the IPD increases by a similar amount (lower panel of Figure 2.7).

In general the model can predict the data for the Gaussian noise masker (right panel of Figure 2.7). It quantitatively predicts the shape of the threshold curve as a function of the IPD. As in the data, the model predicts a smaller CMR for the Gaussian noise than for the multiplied noise masker. The average predicted CMR is 9 dB which is in reasonable agreement with the average measured CMR of about 10 dB.

2.3.2 CMR as a function of the portion of comodulated masker components

Figure 2.8 shows results of the second experiment where the portion of comodulated masker components was varied (left column). In the upper panel detection thresholds are shown as a function of the portion of comodulated masker components for a signal IPD of 0 degrees (circles), 72 degrees (diamonds) and 180 degrees (triangles). Thresholds are expressed as dB relative to the level of the masker centred at the signal frequency. The middle panel shows the CMR, calculated for each value of the IPD as the improvement of thresholds relative to the threshold with a portion of comodulated masker components of 0%. The lower panel indicates the binaural gain, calculated as the difference in thresholds between the diotic condition (IPD=0) and the two dichotic conditions at each value of the portion of comodulated masker components.

The threshold for an IPD of 0 degrees and a portion of comodulated masker components of 0% is 2 dB. Thus, as expected, the same threshold is measured as in the first experiment for a diotic signal masked by an uncorrelated masker. Thresholds decrease monotonically as the portion of comodulated masker components is increased, reaching a minimum value of -12 dB for a completely comodulated masker. A similar decrease is found for the IPDs of 72 and 180 degrees. The maximum threshold is -7 dB for a signal IPD of 72 degrees and -11 dB for 180 degrees. The thresholds are essentially the same as in the first experiment (difference $\leq$ 1 dB). The minimum threshold is obtained for a 100% comodulated masker: -19 dB for an IPD of 72 degrees and -23 dB for 180 degrees. CMR (middle panel of Figure 2.8) increases monotonically as the portion of comodulated masker components increases. The maximum value of the CMR is 14 dB at a portion of comodulated masker components of 100 % for the zero IPD signal and 12 dB for the signals with an IPD of 72 and 180 degrees. Thus the maximum CMR is very similar for the diotic and dichotic signal conditions.

The BMLD (lower panel of Figure 2.8) for an IPD of 72 degrees is at 8 dB for the uncorrelated masker and is almost constant up to a portion of comodulated masker components of 50%. For larger portions of comodulated masker components the BMLD slightly decreases to a value of 7 dB for the comodulated masker. The BMLD of the IPD of 180 degree is an upward shifted version of the BMLD for an IPD of 72 degrees, reaching a value of about 13 dB in the uncorrelated case and 10 dB in the comodulated case. This decrease might be (for the same reasons as for the CMR) due to the variance in the data.

The right column of Figure 2.8 shows simulated results. Simulated thresholds for a portion of comodulated masker compo-

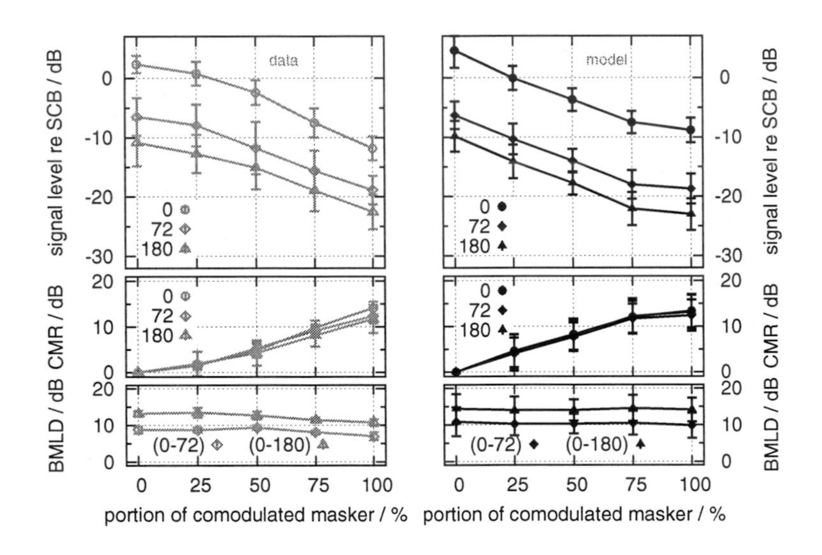

Figure 2.8: Simulated thresholds (black symbols) together with measured thresholds (grey symbols). In the upper row, thresholds are plotted relative to the level of the SCB for IPDs of 0 (circles), 72 (diamonds) and 180 degrees (triangles). In the middle panel, CMR is shown. In the lower panel, the BMLD for the IPD of 72 (diamonds) and 180 degree (triangles) is shown. All curves are plotted as a function of the portion of comodulated masker.

nents of 0% and 100% with an IPD of 0 (circles) and thresholds for a degree of comodulation of 0% and an IPD of 180 degrees (leftmost triangle) correspond to the thresholds shown in Figure 2.5. Thresholds decrease linearly and in parallel as a function of the IPD. The predicted threshold for an IPD of 72 is in good agreement with the experimentally obtained value. The threshold for the same IPD in the comodulated condition is slightly overestimated compared to the experimental data, but well within the interindividual standard deviation. The linear decrease of thresholds deviates from the experimentally obtained course, showing a slower decrease for smaller values of the portion of comodulated masker components. This shows the increase of CMR as shown in the middle panel. CMR increases faster in the simulation than in the experimental data but is well in line with all values of the IPD. The parallel decrease of thresholds in the upper panel leads to the constant value of the BMLD as shown in the lower panel.

2.4 Discussion

The experiments of the present study investigated the influence of interaural phase differences (IPD) on the effect of uncorrelated and coherent temporal masker structure in remote frequency bands. The data showed that the human listeners benefit from across-frequency and binaural information if both cues were present.

Data of the first experiment showed that the monaural across-frequency effect of CMR is not affected by the introduction of interaural disparities through an IPD of the signal. The magnitude of CMR for a signal with no IPD is similar to the CMR found in the literature for a comparable masker paradigm for multiplied noise maskers [Hall *et al.*, 1990] and Gaussian noise maskers [Moore and Shailer, 1991]. The binaural benefit (BMLD) increases independently of the across-frequency cue, i.e. the temporal envelope structure of the noise masker. The magnitude of the BMLD for the antiphasic signal is in qualitative agreement with the BMLD of a previous study for a broadband masker with a similar spectral width and a similar center frequency [Hall and Fernandes, 1983, van de Par and Kohlrausch, 1999].

The overall benefit found in the first experiment is the difference found between the threshold with no additional cues (IPD=0, uncorrelated) compared to the thresholds with both cues, i.e. a non-zero IPD of the signal and a comodulated masker. This is in line with the assumption that the BMLD only depends on the spectral contents of the stimulus and not on the temporal envelope structure of the diotic noise masker. The assumption is further supported by the similarity of the BMLD for the two noise types (i.e. multiplied noise and Gaussian noise).

Only few data exists on comparison of CMR for multiplied noise maskers and Gaussian noise maskers. In the present study, CMR for Gaussian noise maskers is slightly smaller than for multiplied noise maskers. This agrees with previous data on CMR using these two noise types [Moore *et al.*, 1990].

About the same BMLD was measured for all portions of comodulated masker components while the CMR increases monotonically with increasing portion of comodulated masker components. The data point at a portion of comodulated masker components of 100% with no IPD is 2 dB lower than in the corresponding condition in the previous experiment. This deviation expresses the higher maximum value of the CMR for that IPD. It also reflects in the slight decrease in the BMLD in the lower panel since this point is taken as reference for the computation of the BMLD. Given the interindividual standard deviation, there is a very good agreement with the CMR obtained for the other values of the IPD.

These results clearly indicate an independent processing of fine-structure cues and across-frequency envelope cues. The data can not be explained by processing stages aligned in parallel where the best representation of the monaural channels and the binaural channel is used as the model output. Such a parallel processing would predict a dominance of one of the two effects, provided they do not have the same magnitude. Otherwise, the combined effect would be equal to the effect when only one of the cues were present. Thus a parallel model in an equally simple approach as the one used here with a serial alignment of the stages is unable to reproduce the data.

Following the energy detector approach as given in Buus [1985], the addition of the two effects can be modelled as two subsequent improvements of the overall signal-to-noise ratio of the internal representation of the signal in the noise. Such an addition can be implemented by assuming serially aligned processing stages. This hypothesis was further tested by means of a conceptual model of auditory processing.

The across frequency stage was implemented as a wideband inhibition strategy in a neural system. This implementation of a wideband cell within the model does not mimic all characteristics of the response map of an onset neuron that was hypothesized to serve as wideband inhibitors at the level of the cochlear nucleus [e.g., Pressnitzer *et al.*, 2001, Neuert *et al.*, 2004]. For example, wideband cells generally show a gradual decrease in sensitivity towards higher and lower frequencies [e.g., Winter and Palmer, 1995, Jiang *et al.*, 1996]. Such a decrease could be implemented within the current model by assuming a weighted sum of the outputs of frequency channels giving less weight to those channels with centre frequencies well below or above the signal frequency. Such an approach was not implemented in the current study since (i) the current data does not allow to adjust the parameters for the weighting of the frequency channels remote from the signal frequency and (ii) it would increase the number of free parameters without giving further insights into the basic mechanism. The decrease in sensitivity at the signal frequency may be interpreted as a implementation of a onset cell with a bimodal response map, i.e. a second lower threshold peak below the best frequency of the cell [Winter and Palmer, 1995]. An alternative interpretation is that it is a simplified implementation of the combination of a wideband inhibitor cell and a narrowly tuned "disinhibition", i.e. an inhibition of a inhibitory narrowly tuned cell as hypothesized in Meddis *et al.* [2002]. In a recent effective model of CMR a similar strategy was used as in the present study by applying zero weights to channels at and close to the signal frequency for the hypothesized equalization-cancellation process [Piechowiak *et al.*, 2007].

The wideband inhibitor reduced the activation of a narrowband (NB) cell whenever the activation in off-frequency channels is large. The output of the narrowband cell is the input of the binaural stage. The binaural stage was implemented as a simplified version of an equalisation-cancellation mechanism [Durlach, 1963]. In general, model simulations agree with the experimental results in several aspects: (i) CMR is about 11 dB for multiplied noise maskers and about 9 dB for Gaussian noise maskers and constant for all IPDs and (ii) the increase in IPD leads to a monotonic decrease in thresholds for both conditions. The difference between magnitudes of simulated and measured BMLD is smaller than the interindividual standard deviation. The simulations show that already a simplified model of the auditory system is able to quantitatively predict the data when a serial alignment of the stages is assumed. The main goal of the conceptual model was to test the feasibility of the hypotheses of serially-aligned processing stages for CMR and BMLD under the constraint of a minimum set of parameters. While this minimum number of parameters is desirable for an easy interpretation of the model, the downside is a reduced ability of the model to account for all aspects of CMR and BMLD. In order to generalize the model, the model stages for CMR and BMLD might be replaced by more elaborate approaches, leaving the current implementation as a special case of a more general implementation with more parameters. Nevertheless, the serial alignment of the processing stages must be preserved in any implementation to obtain the additivity of CMR and BMLD.

While the results clearly point towards a serial processing, the exact order of the across-frequency and across-ear stage can not be determined on the basis of the current data set. In order to show that the data does not allow the derivation of the order of the processing stages the first experiment was simulated with a modified model, where the the across-ear processing stage preceded the across-frequency processing. The binaural processing was done in the signal channel as well as in the other channels that form the basis for the computation of the across-frequency weighting function. The parameters were the same as used in the original model. A comparison of the predictions of this modified model with those of the original model (shown in Figures 2.6 and 2.7) indicated that both models perform equally well in predicting the data.

With the modified model it is straightforward to predict dichotic CMR, where on-frequency masker and flanking bands are presented to opposite ears since the across frequency stage of the modified model is not sensitive to at which ear the different masker components are presented. Thus, from the dichotic CMR, one may argue that the modified model is better to predict binaural effects in psychoacoustical CMR. However, the model would predict the same dichotic and monaural CMR whereas psychoacoustical data indicate a slightly reduced magnitude of CMR in the dichotic condition [Schooneveldt and Moore, 1989, Hall *et al.*, 1990, Ernst and Verhey, 2006]. A dichotic CMR is not against the idea of across-frequency processing stage prior to the stage responsible for the BMLD. Anatomical and physiological data seems to indicate that wideband-inhibitor cells project contralaterally. Thus, the original model could be extended to account for dichotic CMR including a contralateral projection of the wideband inhibitor [Verhey *et al.*, 2003, Ernst and Verhey, 2006, Ingham *et al.*, 2006], presumably with a reduced amount of inhibition compared to the ipsilateral projection. This modification was not included here for the sake of simplicity. The experimental paradigm used in dichotic CMR studies was not in the scope of the present study - i.e. the combined effect of CMR and BMLD - due to the absence of a BMLD.

Under the assumption that CMR is based on higher order effects after binaural processing like, for example, envelope locking suppression [Nelken *et al.*, 1999], a sensitivity of these neurons to binaural cues would be expected. On the other hand, if CMR processing precedes binaural processing, across-frequency interaction should be expected at the binaural stage.

Whatever the order in the auditory system is, the data imply that also in more complex models, processing of monaural across-frequency cues and binaural disparities has to be implemented independently and aligned in a serial fashion. To find out about the realization of this highly efficient processing in the auditory system, further physiological and psychoacoustical experiments are needed, especially to determine which of the two stages precedes the other.

2.5 Summary and Conclusions

The data showed that human listeners benefit from across-frequency and binaural information when both cues were present. The combined effect is the sum of the gain due to comodulation of the noise masker and the gain due to interaural differences. This additive effect could be found for both masker types investigated. Summation of these two effects leads to an overall benefit of up to 24 dB. A simple model based on the assumption of serially aligned processing stages for across-frequency and binaural processing is able to reproduce the data. The ability of the auditory system to add the masking releases due to each of the cues reflects the high performance of the auditory system under the naturally given constraints of computational capacity and energy demand. The experimental data in connection with the model predictions may stimulate new physiological experiments contributing to the knowledge about the topography and the realization of the signal processing in the auditory system.

Acknowledgments

We would like to thank two anonymous reviewers and the action editor for many helpful comments on a previous version of the manuscript. This work was supported by the Deutsche Forschungsgemeinschaft (International Graduate School for "Neurosensory Science and Systems" GRK 591 and SFB TR31).

3 Combination of masking releases for different center frequencies and masker amplitude statistics[1]

Abstract

Several masking experiments have shown that the auditory system is able to use coherent envelope fluctuations of the masker across frequency within one ear as well as differences in interaural disparity between signal and masker to enhance signal detection. The two effects associated with these abilities are comodulation masking release (CMR) and binaural masking level difference (BMLD). The aim of the present study was to investigate the combination of CMR and BMLD. Thresholds for detecting a sinusoidal signal were measured in a flanking-band paradigm

[1] Reprinted with permission from:
Epp, B. and Verhey, J.L. **(2009)** "Combination of masking releases for different center frequencies and masker amplitude statistics." J. Acoust. Soc. Am. 126 (5) 2479-2489. © 2009, Acoustical Society of America.

at three different signal frequencies. The masker was presented diotically and various interaural phase differences (IPD) of the signal were used. The masker components were either multiplied or Gaussian narrowband noises. In addition, a transposed stimulus was used to increase the BMLD at a high signal frequency. For all frequencies and masker conditions, thresholds decreased as the signal IPD increased and were lower when the masker components were comodulated. The data show an addition of the monaural and binaural masking releases in decibels when masker conditions with and without comodulation and the same spectrum were compared.

3.1 Introduction

An important task of the auditory system in natural acoustical environments is to segregate sounds from different sound sources. It is generally assumed that the auditory system uses monaural cues, such as the coherent envelope fluctuations across frequency – a characteristic of many natural sounds [Nelken *et al.*, 1999] – as well as binaural cues to separate the different sound sources. The aim of the present study was to investigate the ability of the auditory system to combine binaural and monaural cues in a psychoacoustical masking paradigm.

One psychoacoustical phenomenon associated with the ability to use monaural across-frequency cues is comodulation masking release (CMR). Comodulation masking release is the effect that the detectability of a sinusoid masked by a narrowband noise centered at the signal frequency (signal-centered band) can be improved by additional masker bands at spectrally distal positions (commonly referred to as flanking bands), but only if the signal-centered band and the flanking bands show coherent envelope fluctuations, i.e. are comodulated [Hall *et al.*, 1984, Verhey *et al.*, 2003, for a review]. For the flanking-band paradigm, the magnitude of CMR is either calculated as the difference in thresholds with the signal-centered band only (reference condition, RF) and the threshold obtained by addition of comodulated flanking bands (comodulated condition, CM), or is defined as the benefit due to comodulated noise bands compared to noise bands having uncorrelated intensity fluctuations (uncorrelated condition, UN). In the following, the former CMR will be referred to as CMR(RF-CM) and the latter will be referred to as CMR(UN-CM). CMR has been shown to depend on center frequency, number, spectral width and level of the flanking bands, and the statistics of the masker [Schooneveldt and Moore, 1987, Hall *et al.*, 1990, Moore *et al.*, 1990, Verhey *et al.*, 2007, Epp and Verhey, 2009]. CMR tends to increase with signal frequency and with the number of flanking bands. In addition, CMR depends on the masker envelope distribution [Eddins, 2001, Epp and Verhey, 2009]. For spectral configurations with flanking bands close to the signal frequency, it has been suggested that part of the CMR is due to processing the output of one auditory filter [e.g. McFadden, 1986, Schooneveldt and Moore, 1987, Piechowiak *et al.*, 2007]. It was shown by Verhey *et al.* [1999] that a model of the auditory system can predict CMR in one type of CMR experiment, the bandwidening experiment, by exclusively processing the information at the output of the auditory filter centered at the signal frequency. In bandwidening CMR experiments, flanking bands are added implicitly by broadening the masker centered at the signal frequency. Piechowiak *et al.* [2007] showed that the model proposed by Verhey *et al.* [1999] also predicts the CMR in flanking-band CMR experiments, i.e. where a flanking band was added at a spectrally distal position, if a moderate spectral distance to the signal-centered band was used (less than 30% of the signal frequency). For a flanking band more distal to the signal frequency, CMR is assumed to be the result of across-channel processing [Cohen, 1991, Verhey *et al.*, 2003], at least if the signal-centered band and the flanking band have the same level. However, Ernst and Verhey [2006] showed that, for large level differences between the signal-centered band and the flanking band, part of the CMR might be due to suppression at a cochlear level, even when the flanking band center frequency is several octaves below the signal frequency.

The auditory system is also able to use interaural disparities in either the signal or the masker to improve the detectability of the masked signal [Hirsh, 1948, Licklider, 1948]. This effect is referred to as the binaural masking level difference (BMLD) [Jeffress *et al.*, 1956, van de Par and Kohlrausch, 1999]. The BMLD depends, among other things, on the bandwidth of the masker and the signal frequency [Hirsh, 1948, Zurek and

Durlach, 1987, van de Par and Kohlrausch, 1999]. The BMLD decreases as the masker bandwidth increases and has a tendency to decrease with the signal frequency. The question whether the auditory system is able to combine the two cues comodulation and interaural disparities to increase the efficiency of signal detection in noise was addressed by Hall *et al.* [1988, 2006] and Cohen [1991]. Hall *et al.* [1988] investigated CMR when the 500-Hz pure tone signal was masked by a narrow band of noise alone and when a comodulated flanking band was added at a center frequency of 400 Hz. The CMR was found to be larger in conditions with diotic stimulation than in conditions where an antiphasic signal was masked by a diotic noise. All of the subjects showed a reduced or no benefit due to the addition of a comodulated masker band. About half of the subjects could benefit from the comodulated flanking band, but only for the smallest bandwidth of the noise. For the largest bandwidth used in their study, Hall *et al.* [1988] concluded that the auditory system does not seem to be able to benefit from across-frequency information in a dichotic listening condition. Cohen [1991] measured thresholds for a 700-Hz pure tone signal masked by a narrowband masker centered at the signal frequency in the presence and absence of an additional narrowband masker centered at 600 Hz. The stimuli were presented diotically (N_0S_0) and dichotically with an interaurally inverted signal (N_0S_π). They found a reduction of both CMR(RF-CM) and CMR(UN-CM) for the dichotic listening condition compared to the diotic condition. CMR(RF-CM) vanished but a small benefit was observed in the presence of a comodulated flanking band compared to an uncorrelated flanking band. As a small CMR(UN-CM) was found, Cohen [1991] concluded that CMR and BMLD "are, to some extent, additive". In a more recent study of Hall *et al.* [2006], the combination of CMR and BMLD was investigated by changing the interaural correlation of the masker. The stimulus used in their study was a 500-Hz pure tone masked by several narrow bands of noise with a spectral distance of 100 Hz. They found no consistent binaural CMR, i.e. large individual differences and, on average, only a small enhancement of binaural detection due to the presence of the comodulated flanking band.

Schooneveldt and Moore [1989] also investigated the combination of CMR and BMLD using various frequency separations of signal-centered band and flanking band and various monaural and binaural presentations of the masker and signal. The binaural benefit was quantified by comparison of conditions with diotic noise and monaural signal presentation (N_0S_m) and conditions with diotic noise and diotic signal presentation (N_0S_0). They hypothesized sequential processes underlying CMR and BMLD. This study is not directly comparable to the other studies because they investigated the combination of CMR and BMLD only for a binaural gain due to a comparison of a monaural versus diotic signal presentation. In a recent study, Epp and Verhey [2009] showed that data from combined CMR(UN-CM) and BMLD paradigm at 700 Hz using various interaural signal phase differences (IPD) in combination with diotically presented uncorrelated and comodulated masker conditions can be explained using a model with serial alignment of across-frequency and across-ear processing stages. This result supports the hypothesis of Schooneveldt and Moore [1989] that the processing stages underlying CMR and BMLD operate sequentially. However, Epp and Verhey [2009] only investigated the combination of CMR(UN-CM) with the BMLD.

The previous studies on the combination of comodulation and IPD suffer from at least one of the following limitations: (i) they used small spectral distances between the masker components, so within-channel mechanisms may have contributed to the CMR to a large extent (ii) they quantified CMR using only one of the two definitions of CMR and (iii) there is only a very limited set of data for comparison of the two single effects CMR and BMLD and their combination.

The present study attempts to overcome the limitations of these previous studies by measuring the thresholds for detecting a signal in the reference, uncorrelated and comodulated conditions for (i) larger spectral distances between the components, (ii) various interaural phase differences of the signal, and (iii) different signal frequencies. The combined masking release was measured for two different noise types which have commonly been used in CMR experiments. The use of the two noise types facilitates the comparison to previous data in the literature. These two noise types differ in their envelope amplitude distributions and, thus, may provide insights into the mechanisms underlying the across-frequency and across-ear processing.

3.2 General methods

3.2.1 Procedure

A three-alternative, forced-choice procedure with adaptive signal-level adjustment was used to determine the masked threshold of the sinusoidal signal. The intervals in a trial were separated by gaps of 500 ms. Subjects had to indicate which of the intervals contained the signal. Visual feedback was provided after each response. The signal level was adjusted according to a two-down, one-up rule to estimate the 70.7% point on the psychometric function [Levitt, 1971]. The initial step size was 8 dB. After every second reversal, the step size was halved, until a step size of 1 dB was reached. The run was then continued for another six reversals. The mean level at these last six reversals was used as an estimate of the threshold. The final individual threshold estimate was taken as the mean over four threshold estimates.

3.2.2 Stimuli and apparatus

The signal was a pure tone which was temporally centered in the masker and had a duration of 250 ms, including 50-ms raised-cosine ramps at onset and offset. The masker duration was 500 ms, including 50-ms raised-cosine ramps at onset and offset. The masker consisted of one or five noise bands. Each noise band had a bandwidth of 24 Hz and a level of 60 dB SPL. One noise band was centered at the signal frequency (signal-centered band, SCB), and the four flanking noise bands were centered at frequencies remote from the signal frequency (flanking bands). The flanking bands were either absent (reference condition, RF), had uncorrelated intensity fluctuations (uncorrelated condition, UN), or had the same intensity fluctuations (comodulated condition, CM) as the signal-centered band. The masker was presented diotically. The signal had an interaural phase difference (IPD) in the range from 0 to 180 degrees. Two types of masking noise were used: multiplied noise and Gaussian Noise.

Multiplied noise masker bands were generated by multiplying a random phase sinusoidal carrier at the desired center frequency by a narrowband noise, which was lowpass filtered at 12 Hz and where the DC component was removed. This procedure mimics the analog realization of multiplied noise with noise generators that produced signals with zero mean value. A similar procedure was used in previous studies [Ernst and Verhey, 2006]. For the reference (RF) and the uncorrelated (UN) conditions, independent realizations of the lowpass noise were used for each masker band. In the comodulated (CM) condition, the same lowpass noise was used for all masker components.

Gaussian-noise bands were generated in the frequency domain by assigning numbers derived from draws of a normally distributed process to the real and complex parts of the desired frequency components in each band. For the RF and the

UN conditions this was done independently for each noise band, while the numbers of a single draw were assigned to all frequency bands for the CM condition. The real part of the subsequent inverse Fast Fourier Transform yielded the desired waveform.

For both noise types, new random numbers were drawn for each interval and each trial. All signals were generated digitally with a sampling frequency of 44100 Hz using MATLAB. Signals were converted to the analog domain (RME ADI-8 DS), amplified (Tucker Davis Technologies HB7) and presented to the listeners in a double-walled sound-attenuating booth via headphones. The type of headphones differed between the experiments and is specified in the corresponding methods section.

3.2.3 Listeners

Nine listeners participated in each experiment, varying in age from 22 to 28 years (one of them being the first author, BE). None of the listeners had any history of hearing difficulties and their audiometric thresholds were 15 dB HL or less in the relevant frequency range from 125 Hz to 8000 Hz. The listeners had at least 2 h experience in experiments on CMR and binaural experiments before collecting the data. The listeners were the same in the second and third experiments. One of the listeners who participated in the second and third experiments also participated in the first experiment (TK).

3.3 Experiment 1 - CMR and BMLD at 700 Hz

3.3.1 Rationale

To facilitate the interpretation of the combined effect of comodulation and interaural disparities, a signal frequency was chosen at which monaural and binaural cues alone were expected to lead to a masking release of a reasonable magnitude [Hall *et al.*, 1990, van de Par and Kohlrausch, 1999]. Interaural disparities were gradually introduced and systematically combined using the signal-centered band alone, additional uncorrelated maskers and comodulated maskers. The resulting overall release from masking for each cue combination was used to interpret the combined effect.

3.3.2 Methods

The signal and the signal-centered band were located at 700 Hz. The flanking bands were located at 300, 400, 1000 and 1100 Hz. The signal IPD was 0° (diotic), 14.4°, 36°, 72°, 144° or 180° (antiphasic). The stimuli were presented using Sennheiser HDA 200 audiometric headphones.

3.3.3 Results

Figure 3.1 shows individual results for the multiplied noise masker. Thresholds for detecting the signal in diotic conditions were highest for the reference (circles) and the uncorrelated (squares) masker conditions and lowest for the comodulated (triangles) condition. The thresholds for all listeners in all conditions decreased with increasing IPD. This means that for all masker conditions, an increase in the BMLD occured as the IPD increased. The magnitude of the maximum BMLD (difference in threshold for the diotic and the antiphasic conditions) differed across the listeners. The maximum BMLD in the multi-band conditions (uncorrelated and comodulated conditions) varied from about 10 dB for listener DW to about 20 dB for listener BE. In the single-band condition (reference condition) the maximum BMLD varied from about 10 dB for listener MK to 28 dB for listener AK. There were also individual differences in the effect of the number of bands on the BMLD. For

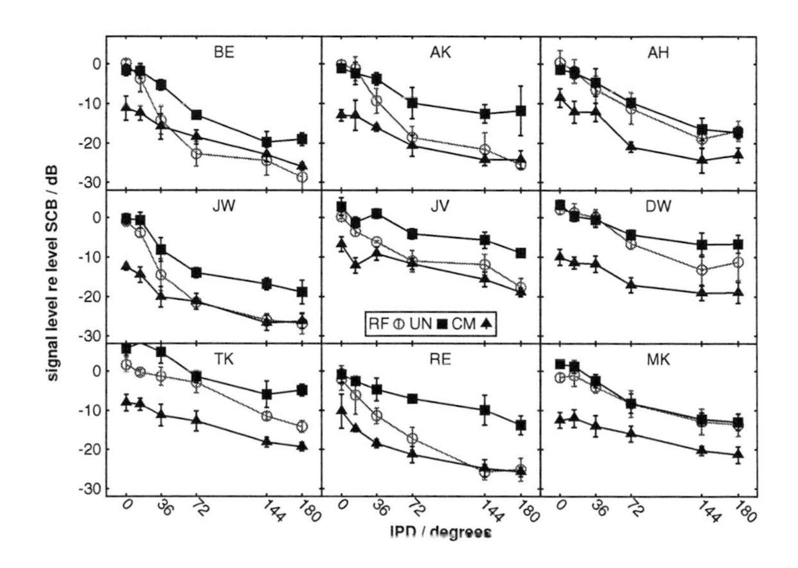

Figure 3.1: Individual data for a signal frequency of 700 Hz and the multiplied noise masker. Mean thresholds are shown over four runs for the reference condition (RF, circles), the uncorrelated condition (UN, squares) and the comodulated condition (CM, triangles). Thresholds are plotted relative to the level of the masker centered at the signal frequency (SCB) as a function of the interaural phase difference (IPD). Error bars indicate ± one standard deviation.

six of the listeners the BMLD was larger for the single-band condition than for the multi-band conditions. Such large individual differences have been reported before [Buss *et al.*, 2007]. Three of the listeners showed no or only minor differences in the BMLD between single-band and multi-band conditions. The diotic CMR(UN-CM) varied from 7 dB (listener AH) to 15 dB (listener TK). For the majority of the listeners, the CMR(UN-CM) in dichotic listening conditions was approximately constant with IPD. A few listeners showed a slight decrease (BE, JW and MK) or increase (TK and RE) of the CMR(UN-CM) with increasing IPD. For most listeners, the CMR (RF-CM) was very similar to the CMR(UN-CM) in the diotic condition but differed from the CMR(UN-CM) in the dichotic conditions. In contrast to CMR(UN-CM), the CMR(RF-CM) decreased for most listeners with increasing IPD. Only listeners AH and MK showed a similar CMR(UN-CM) and CMR(RF-CM).

Figure 3.2 shows mean results for the data shown in Figure 3.1 with interindividual standard deviation. In the upper panel, thresholds are plotted as in Figure 3.1. In the middle panel, the CMR is shown for each value of the IPD. Circles and squares indicate CMR(RF-CM) and CMR(UN-CM), respectively. The lower panel shows the BMLD, i.e. the difference in threshold for each dichotic condition (IPD $\neq$ 0) relative to the threshold for the corresponding diotic condition (IPD = 0). As in the upper panel, circles, squares and triangles indicate BMLDs for the reference, uncorrelated and comodulated conditions, respectively. The average thresholds show a monotonic decrease with increasing IPD. The magnitude of the CMR in the diotic condition was similar for the two definitions of CMR: The diotic CMR(RF-CM) was about 10 dB (middle panel, circles) and the diotic CMR(UN-CM) was about 11 dB (middle panel, triangles). By definition, the BMLD (lower panel) was zero for no IPD. The improvement in thresholds with increasing IPD is reflected in a monotonic increase of the BMLD. The maximum

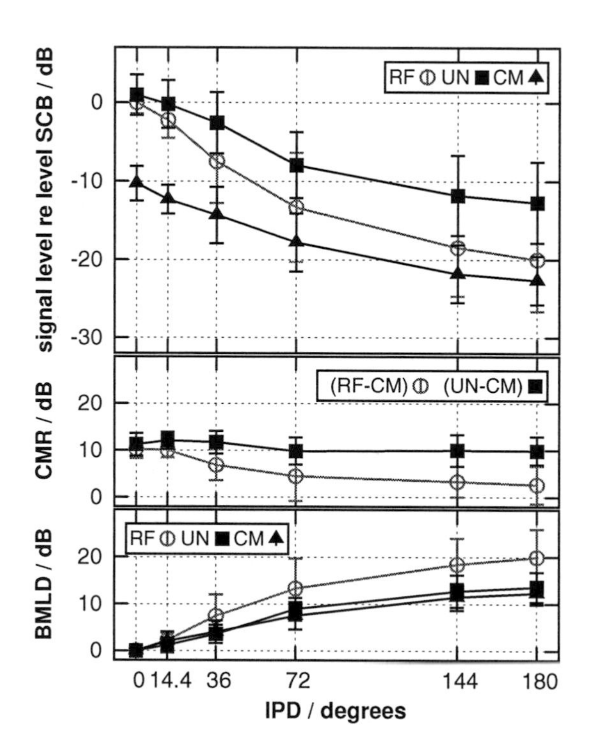

Figure 3.2: Results for the multiplied noise masker. The upper panel shows mean detection thresholds averaged over all listeners for the reference (RF, circles), uncorrelated (UN, squares) and comodulated (CM, triangles) conditions. Thresholds are plotted relative to the level of the masker centered at the signal frequency (SCB). The middle panel shows the average CMR(RF-CM) (circles) and CMR(UN-CM) (squares). In the lower panel, the masking release (BMLD) relative to the diotic condition is shown for RF (circles), UN (squares) and CM (triangles). Error bars indicate ± one standard deviation.

BMLD occurred for an antiphasic signal, and was 20 dB for the reference condition and about 13 dB for the uncorrelated and comodulated conditions. The CMR(UN-CM) was almost constant for the different values of the IPD (deviation of less than 1 dB from the mean value). In contrast, CMR(RF-CM) decreased monotonically as the IPD increased to a minimum value of about 3 dB. The standard deviation of the thresholds (upper panel of Figure 3.2) increased with increasing IPD. In contrast, similar standard deviations of CMR(UN-CM) (middle panel of Figure 3.2) and the BMLD were found for all IPDs for the uncorrelated and comodulated conditions (lower panels of Figure 3.2). This reflects the fact that the individual masked thresholds varied (see Figure 3.1), but the masking releases were very similar for all listeners. The standard deviations were larger for the thresholds in the reference condition than for thresholds in the uncorrelated and comodulated conditions. CMR(RF-CM) also showed larger variability than CMR(UN-CM) and the variability increased when an IPD was introduced.

Figure 3.3 shows mean results for the Gaussian noise masker, plotted in the same manner as in Figure 3.2. Relative to the thresholds for the multiplied noise masker, all thresholds were elevated by about 3 dB. As for the multiplied noise masker, thresholds decreased monotonically with increasing IPD. The CMR(UN-CM) hardly changed (deviation of less than 1 dB from mean value) with increasing IPD while CMR(RF-CM) decreased monotonically with increasing IPD. The maximum BMLD was almost identical to that for the multiplied noise masker (Figure 3.2). For both masker types, the constant CMR(UN-CM) indicates that the auditory system can use comodulation to achieve a masking release in dichotic listening conditions to the same extent as in diotic listening conditions, i.e. the masking released add in decibels. On the other hand, the reduction of CMR(RF-CM), where a single-band condition is compared to a multi-band condition, indicates that, for an antiphasic signal, there is lit-

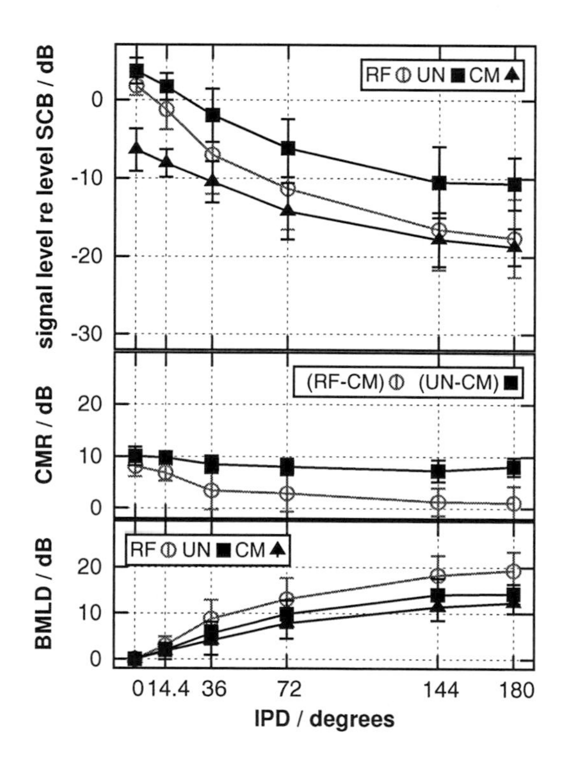

Figure 3.3: As Fig. 3.2 but for the Gaussian noise masker.

tle extra benefit from the comodulated masker bands (circles in middle panel of Figure 3).

3.4 Experiment 2 - CMR and BMLD for higher and lower frequencies

3.4.1 Rationale

To investigate if the additivity of CMR and BMLD in dB also holds for other frequencies, signal frequencies both below and above 700 Hz were used. For lower frequencies, only a small CMR is to be expected [Schooneveldt and Moore, 1987], while a large BMLD should occur [van de Par and Kohlrausch, 1999]. On the other hand a large CMR is to be expected at a higher signal frequency [Schooneveldt and Moore, 1987], while the BMLD is reduced for frequencies above about 1500 Hz due to a gradual loss of fine-structure information [van de Par and Kohlrausch, 1997]. Thus, data at various center frequencies provide insight into the combined effect of comodulation and interaural disparities with different magnitudes of the single effects CMR and BMLD.

3.4.2 Methods

The center frequencies of the masker bands were chosen to have the same ratios of signal frequency and masker band center frequency as those used in the first experiment. The flanking bands were located at 85, 115, 285, and 315 Hz for the 200-Hz signal and at 1285, 1715, 4285, and 4715 Hz for the 3000-Hz signal. The signal IPD was $0°, 72°$or $180°$. The stimulus was presented using Sennheiser HD 650 headphones.

3.4.3 Results

Only mean data are shown since the inter-subject variablility was similar to that for the data of Experiment I. Figure 3.4 shows mean data for the signal frequency of 200 Hz. The left and right panels show results for multiplied and Gaussian noise maskers, respectively. Thresholds are plotted in the upper row. The middle row shows CMR(RF-CM) and CMR(UN-CM) and the lower row shows the BMLD for the reference, uncorrelated and comodulated conditions using the same symbols as in Figure 3.2. Compared to the data obtained at 700 Hz, the CMR was smaller while the BMLD was larger for the multi-band conditions (uncorrelated, comodulated) and slightly smaller for the single-band (reference) condition. For the multiplied noise masker (left panels), the diotic CMR(RF-CM) was about 5 dB and the diotic CMR(UN-CM) was about 8 dB. The thresholds decreased monotonically with increasing IPD. The CMR(RF-CM) and the CMR(UN-CM) hardly varied with the IPD (deviation of less than 1 dB from mean value) and the maximum BMLD was about 17 dB and for all three conditions. For the Gaussian noise masker (right panels), the diotic CMR(RF-CM) was about 0 dB, while CMR(UN-CM) was about 4 dB. The decrease of thresholds with increasing IPD was very similar to that for the multiplied noise masker. The CMR(RF-CM) and CMR(UN-CM) varied little with the IPD (deviation of less than 2 dB and less than 1 dB from mean value, respectively). The maximum BMLD was about 17 dB. The standard deviations of the thresholds (upper panels) increased as the IPD increased. The standard deviations were largest for the reference condition. In contrast, the magnitude of CMR (middle panels) showed only small standard deviations across listeners. The same trend was observed for the data for the 700-Hz signal. The variability of the BMLD (lower panels) was larger than the variability of the CMR, and was largest for the BMLD of the reference condition, reflecting larger individual differences in binaural performance

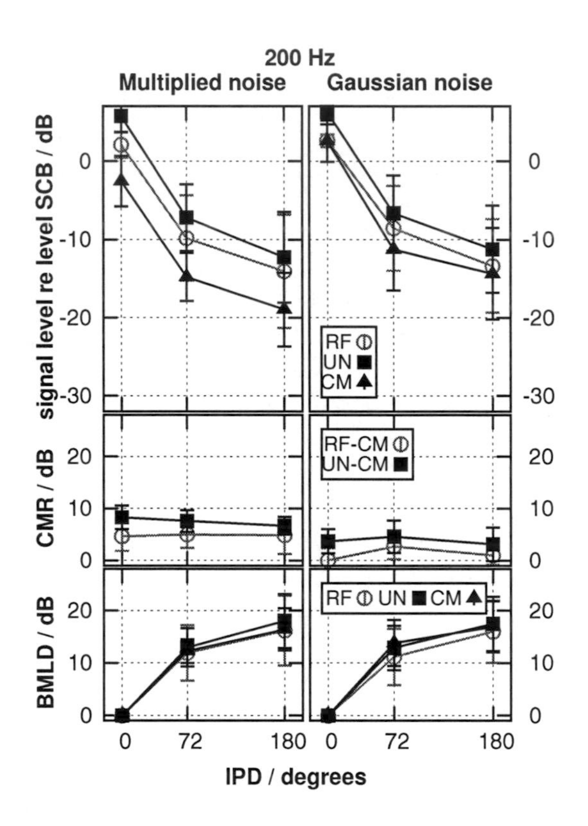

Figure 3.4: Results for a signal frequency of 200 Hz with multiplied and Gaussian noise maskers (left and right panels, respectively). Otherwise, as Figure 3.2.

than in the performance for processing across-frequency cues. Thresholds for the Gaussian noise masker were slightly higher than for the multiplied noise masker and the CMR was slightly smaller, while the BMLD was very similar for the two masker types. The CMR was more affected by the type of noise for the center frequency of 200 Hz than for the center frequency of 700 Hz.

Figure 3.5 shows results for the signal frequency of 3000 Hz for the multiplied (left) and Gaussian (right) noise maskers. The CMR was larger than for the 200-Hz signal and the BMLD was smaller than at 200 and 700 Hz. For the multiplied noise masker, the diotic CMR(RF-CM) was 11 dB and the diotic CMR(UN-CM) was 13 dB. As for the other signal frequencies, thresholds decreased monotonically with increasing IPD. The CMR(RF-CM) and CMR(UN-CM) hardly varied with IPD. The maximum BMLD was about 4 dB for all three conditions. As for the other signal frequencies, thresholds for the Gaussian noise masker were slightly higher than thresholds for the multiplied noise masker and the CMR was smaller by about 5 dB. The diotic CMR(RF-CM) was about 6 dB and the diotic CMR(UN-CM) was 9 dB. The CMR hardly varied with IPD. The maximum BMLD was similar that for the multiplied noise masker (4 dB). Thresholds for the Gaussian noise masker were slightly higher and the CMR was slightly smaller than for the multiplied noise masker. The BMLD was similar for the two noise types and across masker conditions.

The data for signal frequencies of 200 and 700 Hz show the ability to process across-frequency cues in dichotic listening conditions where the BMLD is larger than the CMR. The data for the signal frequency of 3000 Hz also show a constant CMR, independent of the IPD. However, the small magnitude of the BMLD at this center frequency does not allow a clear interpretation of the nature of the combined effect.

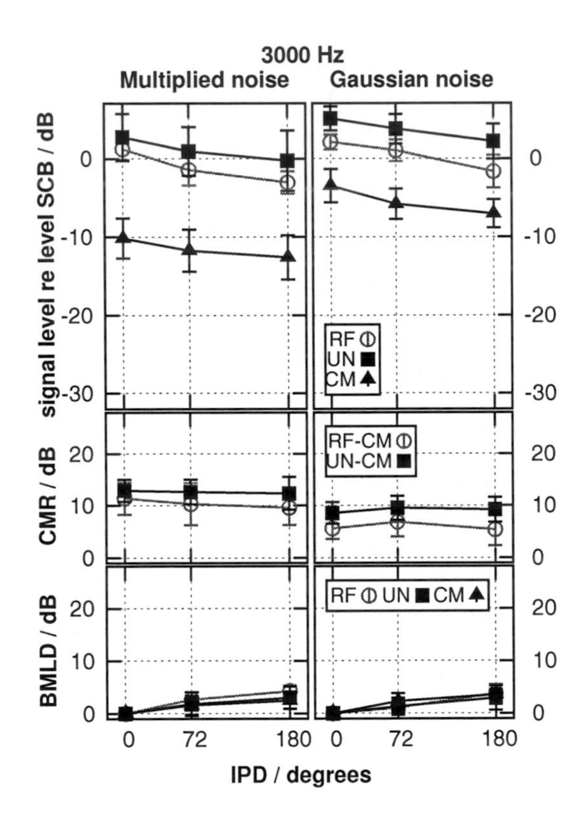

Figure 3.5: As Fig. 3.4, but for a signal frequency of 3000 Hz.

3.5 Experiment 3 - transposed stimulus

3.5.1 Rationale

The maximum BMLD for the signal frequency of 3000 Hz was considerably smaller than the CMR at this frequency. To further investigate how CMR and BMLD combine at high frequencies, a transposed stimulus was used. The transposed stimulus has been shown to increase the BMLD at high signal frequencies [van de Par and Kohlrausch, 1997]. In the present study, the signal-centered band with the signal was transposed from 100 Hz to 3000 Hz. This transposition introduced fluctuations at multiples of 100 Hz into the temporal envelope of the signal-centered band. Thus, even in the comodulated condition, the transposed signal-centered band had a slightly different envelope than the flanking bands. The data of Eddins and Wright [1994] suggest that the auditory system is able to make an across-frequency envelope comparison at different envelope rates simultaneously. In the case where, for example, the modulation is coherent at only one modulation frequency, only the effect for this modulation frequency should be observed. This implies that, for the transposed condition, a CMR for the low rate should still be observed even if the flanking bands and the signal-centered band are different with respect to the high envelope frequency components but are identical in the low envelope frequency components.

3.5.2 Methods

The signal-centered band was generated at a center frequency of 100 Hz and transposed to a center frequency of 3000 Hz. In the target interval, a 100-Hz sinusoidal signal was added to the masker band prior to transposition. The transposition was done by half-wave rectification and lowpass filtering (second order, 500-Hz cut-off frequency) of the low-frequency waveform and

subsequent multiplication with a 3000-Hz carrier.
Multiplied or Gaussian noise flanking masker bands were added at the same frequencies as in the second experiment. In order to avoid spectral overlap of the sidebands introduced by the transposition, the flanking bands were not transposed from low frequencies to high frequencies but were generated as in the experiment for the non-transposed 3000-Hz center frequency. Figure 3.6 shows the time signals (upper panels) and the spectrum (lower panel) of the signal centered band (grey) and the flanking bands (black). There are sidebands around the signal-centered band as a result of the transposition. The flanking bands were centered at 1285, 1715, 4285 and 4715 Hz. The stimulus was presented using HD 650 headphones.

3.5.3 Results

Figure 3.7 shows mean results for the transposed stimulus in the same format as Figure 3.5. The results were comparable to those obtained for the non-transposed 3000 Hz signal. The CMR was slightly smaller and the BMLD was larger for the transposed stimulus than for the non-transposed stimulus. For the multiplied noise masker, the diotic CMR(UN-CM) was about 10 dB and varied little with IPD. The diotic CMR(RF-CM) was about 6 dB and decreased with increasing IPD to 1 dB for an antiphasic signal. The maximum BMLD was larger than for the 3000 Hz condition where the signal-centered band was not transposed (see Sec.3.4). It was 11 dB for the reference condition and about 6 dB for the uncorrelated and comodulated conditions. The diotic CMR(UN-CM) for the Gaussian noise masker was about 5 dB and did not vary with IPD. The diotic CMR(RF-CM) was about 3 dB and reduced with increasing IPD to a negative value for the antiphasic signal. The BMLDs were about the same as for the multiplied noise masker. In contrast to the data where the signal-centered band was not transposed (Fig. 3.5), the standard deviations of the thresholds (upper

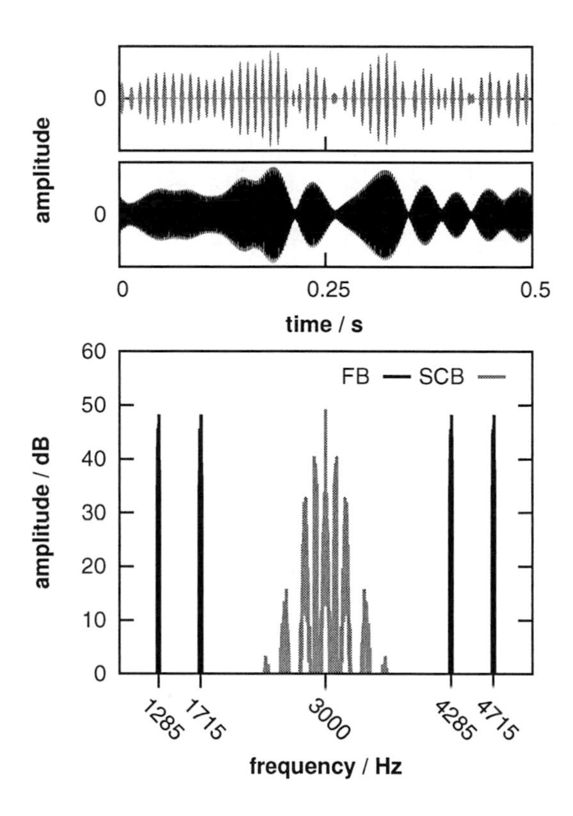

Figure 3.6: Time signal and spectrum of a Gaussian noise masker sample. Upper panel: Time signals of the signal centered band (SCB, grey) and the flanking band (FB, black) centered at 4285 Hz. Lower panel: Spectrum of transposed masker band centered at the signal frequency (SCB) together with additional flanking bands. The SCB shows additional side bands as a result of the transposition, as described by van de Par and Kohlrausch [1997]. The levels of the added flanking bands are well above the levels of the sidebands generated by the transposition of the SCB.

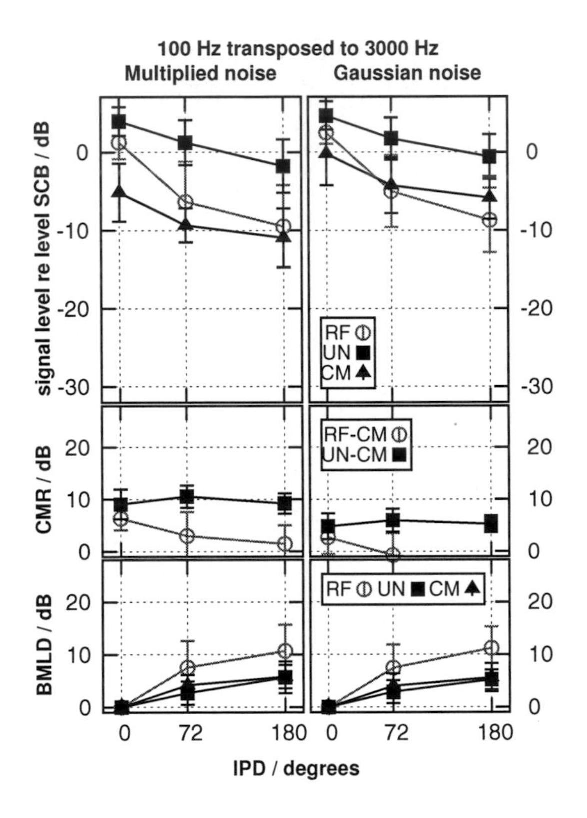

Figure 3.7: As Fig. 3.5, but for the transposed stimulus.

row of Fig. 3.7) indicate large individual differences in masked thresholds. But, as for the data where the signal-centered band was not transposed, the CMR(UN-CM) showed only small variation across listeners. The same holds true for the BMLDs for the uncorrelated and comodulated conditions for both masker types. The standard deviations of the CMR(RF-CM) and of the BMLD for the reference condition increased with increasing IPD.Thresholds for the Gaussian noise masker were slightly higher than thresholds for the multiplied noise masker. The CMR(RF-CM) was smaller for the Gaussian noise masker than for the multiplied noise masker and was even negative for higher values of the IPD. For both masker types, the CMR(UN-CM) did not vary with IPD. In contrast, the BMLDs obtained for the reference, uncorrelated and comodulated conditions were almost identical for the two masker types.

3.6 Discussion

Epp and Verhey [2009] showed that CMR(UN-CM) and BMLD are additive in decibels at a signal frequency of 700 Hz. Their data indicated that the ability to use comodulation cues across frequency was not affected by the introduction of binaural cues and vice versa. The results of the present study support this hypothesis for the 700-Hz frequency for a different set of subjects. In addition, the data show that CMR(UN-CM) and BMLD are additive for signal frequencies of 200 and 3000 Hz. The additivity does not seem to be restricted to a particular signal frequency. Thus, the cues comodulation and interaural phase differences appear to be processed independently in the auditory system. Within the same set of subjects, the additivity is not found for CMR(RF-CM) and BMLD.

3.6.1 Comparison with previous studies focussing on either CMR or BMLD

In general, the results on CMR in the diotic condition (i.e. with no IPD) and those on the BMLD with an antiphasic signal (IPD of 180 degrees) are in good agreement with data found in the literature [Hall *et al.*, 1990, Schooneveldt and Moore, 1989, van de Par and Kohlrausch, 1999].

Hall *et al.* [1990] used 20-Hz-wide multiplied noise masker bands with the same center frequencies as used in the first experiment of the present study and a signal frequency of 700 Hz. They reported a CMR(RF-CM) of about 11 dB. This is in good agreement with the CMR(RF-CM) found in the present study (10 dB). Data from Schooneveldt and Moore [1989] with one flanking band only indicated an increase of diotic CMR(UN-CM) from 5 dB to 8 dB as the center frequency was increased from 250 to 4000 Hz. A similar effect of center frequency was found in the present study: for a center frequency of 200 Hz, CMR(UN-CM) was 7 dB for the multiplied noise masker and 3 dB for the Gaussian noise masker. For a signal frequency of 3000 Hz, the CMR(UN-CM) was about 6 dB larger than for the 200 Hz signal. The differences in the magnitude of CMR between the present study and the study of Schooneveldt and Moore [1989] are presumably due to the higher number of flanking bands used in the present study.

In agreement with Epp and Verhey [2009], the present data indicate that the thresholds for Gaussian noise maskers are higher than corresponding thresholds for multiplied noise maskers and that the CMR tends to be slightly smaller for Gaussian noise than for multiplied noise. This could be explained in terms of "listening in the valleys" [Buus, 1985], since Gaussian noise has less envelope amplitude values close to zero, at which the instantaneous signal-to-noise ratio is high and might be used to improve detectability.

In the present study, the BMLD for the reference condition and a signal IPD of 180 degrees was about 12 dB larger for center frequencies of 200 and 700 Hz than for the center frequency of 3000 Hz. A similar trend was observed by van de Par and Kohlrausch [1999]. For a diotic 25-Hz wide Gaussian noise masker and an antiphasic signal, they found BMLDs of about 25 dB at 500 Hz, 23 dB at 250 Hz and 8 dB at 4000 Hz[2]. In the present study, the BMLD for the single-band condition with a 3000-Hz signal was increased by about 7 dB when a transposed stimulus was used. This increase agrees with the results of van de Par and Kohlrausch [1997], who used a stimulus centered at 125 Hz with a masker bandwidth of 25 Hz which was transposed to 4000 Hz.

For each signal frequency, the maximum BMLDs were the same for the two multi-band conditions (uncorrelated and comodulated conditions) and did not differ between the two masker types. The maximum BMLDs for the multi-band conditions were smaller than for the single-band masker (reference con-

[2]Note that, for all signal frequencies, the magnitude of BMLD reported in the study of van de Par and Kohlrausch [1999] is larger than in the present study. This may be partly due to individual differences. Only three subjects participated in their study, among them the two authors who are certainly both highly trained in binaural listening tasks. An additional problem for the comparison of broadband data with data of a multi-band paradigm is the difference in spectral content of the masker. The spectral notches in a flanking paradigm lead to differences in the modulation spectrum compared to a masker with a continuous spectrum and the same minimum and maximum frequencies. The comparison with data from the literature using a similar spectral range was included here, since it was the most comparable data set to the masking condition used in the multi-band conditions.

dition). A similar effect was found in previous BMLD studies where a reduced BMLD was found when the masker bandwidth was increased. It was hypothesized by van de Par and Kohlrausch [1999] that this reduction might be due to the hampered ability to use off-frequency filters in such spectrally broad conditions compared to narrowband conditions. In terms of this deterioration, the addition of flanking bands might have the same effect as the broadening of the masker.

For a center frequency of 1000 Hz, van de Par and Kohlrausch [1999] found a reduction of the BMLD from 25 to 16 dB when the masker bandwidth was increased from 25 to 1000 Hz. At 4000 Hz, the reduction was only 2 dB. In line with their data, the difference in BMLD between the single-band condition and the multi-band conditions in the present study was about 6 dB for 700 Hz and less than 2 dB at 3000 Hz. Note that the difference was also 6 dB at a signal frequency of 3000 Hz when transposed signals were used.

For the 200-Hz signal frequency, hardly any difference in the BMLD was observed between the single-band and multi-band conditions, whereas for a comparable signal frequency (250 Hz), van de Par and Kohlrausch [1999] found a substantial decrease in the BMLD as the bandwidth increased. The difference between the studies may reflect individual differences. The individual data for the signal frequency of 700 Hz (Fig. 3.1) show that at least two subjects (MK, AH) had a similar BMLD for the multi-band and single-band conditions, although the majority of the listeners showed a larger BMLD for the single-band (reference) condition.

In general, the highest variability in the thresholds and the BMLDs was found for the narrowband reference condition. This finding is in agreement with results of Buss *et al.* [2007]. On the basis of their results, Buss *et al.* [2007] argued that the large individual differences in binaural performance for narrowband maskers may indicate that there are good and poor binaural listeners in experiments using narrowband noise as masker.

Note that the BMLD, in contrast to the CMR, does not show any dependence on the statistics of the masker. The difference in CMR between multiplied and Gaussian maskers might be related to the smaller modulation depth of Gaussian noise compared to multiplied noise. In terms of a "listening in the valleys" approach [Buus, 1985], a smaller modulation depth leads to a lower signal-to-noise ratio in the valleys and consequently to increased thresholds. Such an effect of modulation depth had been observed by Verhey *et al.* [1999] in a bandwidening type of CMR experiment and by Eddins [2001] for a flanking band type of experiment using low-noise noise.

For the BMLD, the data may be understood in terms of an "equalization-cancellation mechanism" [Durlach, 1963]. Cancellation of a diotic masker is independent of the actual statistical properties of the masker under the assumption that the cancellation is performed in an optimal way or that the error of the cancellation is negligible.

3.6.2 Comparison to previous data on the combined effect of interaural differences and comodulation

There are only a few studies on CMR with dichotic signals. These studies differ in their results and their interpretation of the data. Using a flanking-band paradigm, Hall *et al.* [1988] found a decrease in CMR(RF-CM) in dichotic conditions compared to diotic conditions: On average, CMR(RF-CM) was 7 dB in the diotic condition and 2 dB in the dichotic condition. While half of the subjects had a small CMR(RF-CM), the other listeners did not benefit from the additional comodulated masker band. They concluded that there are large individual differences in benefit due to an added comodulated masker band in dichotic listening conditions. Cohen [1991] found a diotic CMR(RF-CM) of about 6 dB and a negligible CMR(RF-CM) when the signal was presented with an IPD of 180 degrees. The CMR(UN-CM) decreased from 9 dB to 4 dB when the signal phase was changed from 0 to 180 degrees. Thus a residual CMR was observed when the uncorrelated condition was used as a reference. On the basis of this result, Cohen [1991] concluded that CMR and BMLD are additive to some extent.

In a recent study, Hall *et al.* [2006] found a large influence of the definition of CMR on the difference between the CMR for a diotic and a dichotic signal. When the signal IPD was changed from 0 to 180 degrees, the CMR(RF-CM) decreased from 12 dB to less than 2 dB while the CMR(UN-CM) decreased from 12 dB to about 5 dB. Compared to the present study, the diotic CMR(UN-CM) is larger in the study of Hall *et al.* [2006] while the dichotic CMR(UN-CM) is smaller. This might be due to differences in the experimental parameters like e.g. the spectral distance between the flanking bands and the signal-centered band. Due to this smaller spectral distance, it is more likely that within-channel CMR contributed to the masking release, leading to a larger CMR in the diotic condition. Hall *et al.* [2006] used experimentally derived psychometric functions in connection with a signal-detection-theory approach to investigate the mechanism of the thresholds obtained for a comodulated masker and dichotic signal presentation. They showed that their data could not be modeled using addition of d' within an integration model [Green and Swets, 1988]. They concluded that the combined effect of comodulation and binaural cues is larger than would be expected from a simple addition of the d' values.

A similar influence of the definition of CMR was observed in the present study for a comparable signal frequency (700 Hz). The CMR(RF-CM) decreased from about 8 to 10 dB (depending on the masker type) for the diotic signal to 1 to 3 dB for the dichotic signal. A similar decrease was found for the transposed stimuli. In contrast, CMR(UN-CM) showed a reduction in the masking release of less than 1 dB. The reduced CMR(RF-CM) in the dichotic condition found by Hall *et al.* [2006] is in qualitative agreement with the reduced CMR(RF-CM) found in the present study for all dichotic conditions. But there is one clear difference: while Hall *et al.* [2006] observed a decrease in CMR(UN-CM) when changing the interaural signal phase from 0 to 180 degrees, the results of the present study indicate that the CMR(UN-CM) is independent of the IPD, i.e the two masking releases add. This qualitative difference may be due to the differences in the experimental parameters. It is likely that the CMR found by Hall *et al.* [2006] was mainly due to within-channel processes, since they used a small spectral distance between the signal-centered band and the nearest flanking band[3]. In contrast, the present study used a three times larger spectral distance. McFadden [1986] and Schooneveldt and Moore [1987] suggested that a large part of CMR in a flanking band experiment with close proximity of the on- and off-frequency masker components may be due to within-channel cues rather than reflecting an across-channel effect. In line with that hypothesis, Piechowiak *et al.* [2007] showed that, for the spectral distance (relative to the signal frequency) comparable to the one used in Hall *et al.* [2006], most of the CMR(UN-CM) can be predicted by a within-channel model. For the spectral distance used in the present study, the model predicted a negligible CMR. This indicates that the CMR found in the present study was mainly due to across-channel processes. Thus, as CMR(UN-CM) was found to be constant with the IPD for the spectral distance used

[3]In one experiment, Hall *et al.* [2006] also used a larger spectral distance between the signal-centered band and the flanking bands. However, for this spectral distance they did not measure thresholds for the uncorrelated condition.

in the present study, CMR(UN-CM) may only be independent of the IPD if it mainly results from across-channel processes. A hypothesis that can be derived from this result is that, in conditions where the diotic CMR is due to a combination of within- and across-channel contributions, the dichotic CMR only reflects the part of CMR that is due to across-channel processes.

3.6.3 Implications for the underlying mechanism

The influence of the definition of the CMR on the effect of IPD complicates the interpretation of the nature of the combination of monaural across-channel cues and binaural cues. Thus, before discussing the possible underlying mechanism it is important to understand the effect of the definition of CMR on the results. In this context it is interesting to reconsider the difference in BMLD for the different masking conditions.

The data of Hall *et al.* [1988] showed that the BMLD with no flanker present had an average value of about 22 dB, whereas the BMLD for the comodulated condition was only 17 dB. A similar reduction of the BMLD was found by Cohen [1991], Hall *et al.* [2006] and in the present study.

The main difference between the reference and comodulated conditions, apart from the comodulation, is the spectrum of the masker. As mentioned before, an increase in the number of spectral components might have a similar effect on the BMLD as an increase in bandwidth which decreases the BMLD. Thus the difference in the BMLD for the spectrally broad (uncorrelated, comodulated) and spectrally narrow conditions (reference) might simply reflect the hampered ability to use off-frequency filters as proposed as an explanation for the bandwidth dependence of the BMLD [van de Par and Kohlrausch, 1999]. Hence, the reduction of CMR(RF-CM) might be the result of two effects: A reduction in threshold due to comodulation and an increase in threshold due to a change in the masker spectrum. The results for the uncorrelated condition of the present study are consistent with the hypothesis that the reduction of the dichotic compared to the diotic CMR(RF-CM) reflects the reduced ability of the binaural system to use off-frequency information in the comodulated condition compared to the reference condition rather than the reduced ability of the auditory system to use across-frequency information.

This hypothesis is supported by the data of Hall *et al.* [2006] using different interaural correlation of the masker. Reducing the interaural correlation to 0.95 abolishes the difference in the BMLD for the reference and the uncorrelated conditions, since for this reduced interaural correlation beneficial across-channel processes can no longer be used by the auditory system [van de Par and Kohlrausch, 1999]. As a consequence, the same dichotic CMR is obtained with the two definitions of CMR. The smaller dichotic CMR(UN-CM) reported by Hall *et al.* [2006] is presumably a consequence of the large contribution of within-channel cues to the CMR in their study (see above). The results of the present study indicate that the single-band (reference, RF) condition is not an appropriate reference to study the combined effect of monaural and binaural cues due to (i) large interindividual difference and (ii) effects due to differences in spectra changing the ability to use off-frequency information in dichotic conditions.

The present data for CMR(UN-CM) indicate that the two masking releases are additive in dB, i.e. the overall masking release is the sum of the CMR(UN-CM) and the BMLD in decibels at each value of the IPD. The additivity of across-channel CMR and BMLD might provide insights into the topographic organization of the processing stages involved in CMR and BMLD processing: Epp and Verhey [2009] showed that a conceptual model based on serial processing stages was able to account for the data from a combined CMR(UN-CM) and BMLD experiment at a signal frequency of 700 Hz. The data of Hall *et al.* [1988],

Cohen [1991] and Hall *et al.* [2006] show a combined effect less than a summation, which is presumably due to the influence of within-channel cues and the comparison of conditions with different spectra, i.e. CMR(RF-CM) in dichotic CMR paradigms. This difference in the ability to combine CMR and BMLD with small and large spectral distance of the flanking bands from the signal frequency may serve to disentangle within-channel and across-channel contributions in CMR paradigms.

The data for the transposed stimuli are in line with the hypothesis that comodulation and interaural phase difference cues are processed independently in the auditory system. The results show a similar magnitude of the CMR as for the data of the non-transposed 3000-Hz stimulus. The similarity in the CMR suggests that the processing stage of CMR is unaffected by the transposition, presumably since the across-frequency envelope cues are preserved. In contrast, the BMLD is increased for the transposed stimuli since this procedure introduces interaural time delay cues that would normally only be available at low signal frequencies.

The invariance of CMR(UN-CM) with the IPD at various magnitudes of the CMR(UN-CM) and the BMLD has two implications for the underlying mechanisms. First, the performance of the processing stage which uses either comodulation or IPD is not affected by the other cue, i.e. these two cues seem to be processed independently in the auditory system. Second, the additivity of CMR(UN-CM) and BMLD in dB can be interpreted as a progressive improvement of the internal representation of the masked signal along the ascending auditory pathway. Such an improvement could be realized as a serial alignment of the underlying processing stages.

3.7 Summary & Conclusions

We investigated the ability of the auditory system to benefit from processing of across-frequency and across-ear cues simultaneously using CMR experiments with flanking bands and various interaural phase differences of the signal. The results show:

(i) CMR(RF-CM) and CMR(UN-CM) were similar in a diotic condition, but differed in dichotic conditions. While CMR(RF-CM) decreased with the introduction of an interaural signal phase difference, CMR(UN-CM) was almost unaffected by an interaural phase difference of the signal.

(ii) The decrease of CMR(RF-CM) in dichotic listening conditions may reflect the reduced ability to use off-frequency filters to process interaural disparities (which was hypothesized to explain the effect of bandwidth on the BMLD) rather than a reduced ability to process comodulation in a dichotic listening condition. Thus, the reference (RF) condition of the CMR flanking-band paradigm might be a problematic reference if the influence of binaural cues is investigated since the masker spectrum for the reference condition is different from that for the other two conditions (UN, CM). In addition, CMR(RF-CM) strongly depends on the IPD and reference thresholds show a large variability across subjects. This also hampers the interpretation of the combination of monaural across-channel and binaural cues.

(iii) The comparison with the uncorrelated condition does not suffer from the interfering effect of the width of the spectrum covered by masker components on the magnitude of the BMLD. For this comparison, a summation of the benefit due to comodulation (CMR(UN-CM)) and the benefit due to an interaural phase difference (BMLD) was found.

The addition was also found for a transposed stimulus. This indicates that the uncorrelated condition is a less problematic reference for quantification of the masking release due to comodulation in dichotic listening conditions. CMR(UN-CM) is not dependent on the IPD and shows only small variability across subjects.

(iv) The additivity of across-channel CMR and BMLD holds true for multiplied noise as well as for Gaussian noise, i.e. the additivity of CMR and BMLD is independent of the envelope amplitude distribution of the masking noise used in the present study.

(v) The same CMR(UN-CM) for all IPDs suggests independent processing of CMR and BMLD and supports the hypothesis of Schooneveldt and Moore [1989] and Epp and Verhey [2009] of a serial arrangement of the processing stages underlying CMR and BMLD.

Ackknowledgments

We would like to thank the associate editor Brian Moore, Joe Hall and one anonymous reviewer for many helpful comments on a previous version of this manuscript. This work was supported by the Deutsche Forschungsgemeinschaft (International Graduate School for "Neurosensory Science and Systems" GRK 591 and SFB TR31).

4 Modeling cochlear mechanics: Interrelation between cochlea mechanics and psychoacoustics[1]

Abstract

A model of the cochlea was used to bridge the gap between model approaches commonly used to investigate phenomena related to otoacoustic emissions and more filter-based model approaches often used in psychoacoustics. In the present study, a nonlinear and active one-dimensional transmissionline model was developed that accounts for several aspects of physiological data with a single fixed parameter set. The model shows plausible excitation patterns and an input-output function similar to the linear-compressive-linear function as hypothesized in psychoa-

[1] Accepted for publication in the "Journal of the Acoustical Society of America" as:
Epp, B., Verhey, J.L., Mauermann, M. "Modeling cochlear mechanics: Interrelation between cochlea mechanics and psychoacoustics."

coustics. The model shows realistic results in a two-tone suppression paradigm and a plausible growth function of the $2f_1$*-*f_2 *component of distortion product otoacoustic emissions. Finestructure was found in simulated stimulus-frequency otoacoustic emissions (SFOAE) with realistic levels and rapid phase rotation. A plausible "threshold in quiet" including finestructure and spontaneous otoacoustic emissions (SOAE) could be simulated. It is further shown that psychoacoustical data of modulation detection near threshold can be explained by the mechanical dynamics of the modeled healthy cochlea. It is discussed that such a model can be used to investigate the representation of acoustic signals in healthy and impaired cochleae at this early stage of the auditory pathway for both, physiological as well as psychoacoustical paradigms.*

4.1 Introduction

The cochlea plays a crucial role in auditory signal processing. Before being processed at higher stages of the auditory pathway, all acoustical signals must pass the mechanical preprocessing of the cochlea including frequency decomposition and the mechano-electrical transduction in the inner hair cells. Information that is lost at this stage will not be available for further processing at higher stages of the auditory pathway. In order to improve detection of low-intensity sounds, the cochlea shows active amplification at low levels [Davis, 1983, deBoer, 1983, Hudspeth, 2008]. An increase in the dynamic range is accomplished by compression of the range of incoming intensities [Ruggero *et al.*, 1997]. Theoretical arguments show that the demand of a large frequency range under the boundary conditions of little space and high frequency selectivity is solved in an optimal manner by the mechanics in the cochlea [Zweig *et al.*, 1976]. Impairment of one of these mechanisms has negative consequences for perceptual performance, ranging from the disability to detect certain frequencies or sounds at low levels up to a degradation of speech intelligibility and the ability to form auditory objects in complex acoustical environments. Hence, a detailed understanding of the preprocessing and the influence of impairments at the level of the cochlea is crucial for our understanding of processing mechanisms at higher stages of the auditory pathway.
Peripheral preprocessing in psychoacoustical models is mainly realized in functional, not physiologically motivated physical models. Such functional models are commonly implemented in the form of digital and independent filters or filterbanks like the linear gammatone filter bank [e.g. used in Dau *et al.*, 1997] or, accounting for level effects, the gammachirp filterbank [Irino and Patterson, 2006]. An approach to include suppression in such a functional model of the cochlea is realized by the dual-resonance-nonlinear(DRNL)-filter [Meddis *et al.*, 2001, Plack *et al.*, 2002]

and was used as a frontend in some auditory models [Ernst and Verhey, 2006, Jepsen *et al.*, 2008]. An important drawback of these models is that the system is simulated without interaction between frequency channels, such that experimentally relevant effects like otoacoustic emissions or effects of finestructure within the cochlea can not be accounted for. On the other hand, physical models that are commonly used to investigate otoacoustic emissions describe the cochlea in more detail, implicitly including the aspects accounted for by functional models. Hence, a physical model that is able to account for physiological and psychophysical data simultaneously can be of great value to investigate the influence of cochlea mechanics and cochlea damage on perception.

There are different ways to evaluate the condition of the inner ear. Probably the most common experiment is to measure the smallest intensity necessary to detect a pure tone at discrete frequencies over a broad frequency range, i.e. an audiogram. When an audiogram is measured with a high frequency resolution, the audiograms of normal hearing listeners often show quasi-periodic fluctuations, referred to as threshold microstructure [Elliott, 1958, Thomas, 1975, Long and Tubis, 1988] or hearing threshold finestructure [Mauermann *et al.*, 2004, Heise *et al.*, 2008] (in the following the term threshold finestructure will be used). Spontaneous otoacoustic emissions (SOAE) are often found in regions of high sensitivity, i.e. in microstructure minima, indicating a common underlying mechanism [Schloth and Zwicker, 1983, Zwicker and Schloth, 1984]. Finestructure is highly sensitive to the application of ototoxic drugs [Long and Tubis, 1988]. Since several studies show an anticorrelation between threshold finestructure and the degree of hearing loss [Long and Tubis, 1988, Horst *et al.*, 2003] it was proposed by Heise *et al.* [2008] that the presence of finestructure might be used as a sensitive measure of cochlea condition. It was argued that, in order to get finestructure, the excitation on the basilar membrane (excitation pattern) due to a pure tone must be sharply tuned and be both, "broad and tall". In addition, small irregularities in the properties of the basilar membrane, so-called "roughness" must be present [Zweig and Shera, 1995]. These broad and tall excitation patterns were found in healthy preparations of cochleae in various species [Rhode, 1971, Robles and Ruggero, 2001], but only when the animal was alive. The sharp tuning vanishes within minutes after death of the animal [Nuttall and Dolan, 1996]. An active process was proposed to be present in the mechanics of the cochlea [Davis, 1983] and it was shown that such an active process is a necessary condition in order to obtain excitation patterns in a model that match physiological data [deBoer, 1983]. Physiological data also indicate that the sharp tuning gradually decreases from the basal end of the cochlea to the apical end [Cooper and Rhode, 1997]. Threshold finestructure also has consequences on the performance of listeners in psychoacoustic tasks. Cohen [1982] showed that temporal integration of pure tones is different for maxima and minima of threshold finestructure. In the context of loudness perception, Mauermann *et al.* [2004] showed that finestructure can also be found in iso-loudness contours up to 40 phon. Heise *et al.* [2009b] showed that performance to detect amplitude modulation of a sinusoidally amplitude modulated tone of low intensity depends on the relative position of the spectral stimulus components to the threshold finestructure of the listener. Heise *et al.* [2009a] hypothesized that the differences in modulation detection performance can be explained at the level of the cochlea. They discussed three possible explanations of the fine structure on modulation perception. The first explanation was that a beating between spontaneous otoacoustic emissions (or more general cochlear resonances) and the stimulus (monaural diplacusis) may hamper the detection performance. Another possible explanation was that the effective modulation depth of the stimulus is reduced by cochlear resonances that synchronize to the carrier component at frequencies of finestructure minima. The third explanation is based on the hypothesis that the fine

structure can be interpreted as a frequency specific gain. This comb filter-like gain function would be the inverse of the fine structure. Heise *et al.* [2009a] concluded on the basis of their results that these explanations can not account for the whole effect, but that the hypothesis of synchronization of cochlea resonances is the most adequate.
Another way to investigate the condition of the inner ear is to measure signals that are generated and emitted from the inner ear, so-called otoacoustic emissions (OAE) [Kemp, 1978]. In a healthy cochlea, finestructure effects can also be found in various types of otoacoustic emissions like distortion products of otoacoustic emissions (DPOAE, or combination tones) [e.g. Mauermann *et al.*, 1999a] or stimulus frequency otoacoustic emissions (SFOAE) [e.g. Zweig and Shera, 1995]. Combination tones are the result of the nonlinearity of the cochlea, as also shown in physiological studies [Rhode, 1971, Robles and Ruggero, 2001, for a review]. For DPOAE, the finestructure is thought to be the result of a superposition waves emitted by two sources. The first (wave-fixed) source is region where the excitation patterns of the primary tones overlap as the source where the distortion products are produced. The second source (place-fixed source) is the place where the wave is reflected at irregularities in the mechanical properties of the cochlea. Part of the energy of this multiply reflected wave is emitted into the ear canal [Shera and Zweig, 1993]. Another consequence of the nonlinearity of the cochlea is the effect of two-tone suppression, i.e. the reduced response to one tone when a second tone at a different frequency is present [e.g. Duifhuis, 1980, Ruggero *et al.*, 1992]. One way to characterize the nonlinearity are input-output (I/O) curves, measured either physiologically by measuring the response to an external tone at a fixed place on the basilar membrane [Rhode, 1971] or derived from psychoacoustical data [Nelson *et al.*, 2001]. The exact shape of the measured nonlinearity differs. While there is general agreement that the healthy cochlea shows a compressive growth in its response to levels between 20 and 100 dB SPL, it

is still unclear if and at which levels the cochlea linearizes for low and high levels. Some data indicate that at the base of the chinchilla cochlea the growth is compressive from very low levels up to levels of at least 100 dB SPL in the vicinity of the position of maximum response for a low level tone (characteristic place, CP), if the preparation is healthy [Ruggero *et al.*, 1997]. For frequencies well above and below CP, the growth is approximately linear. Data derived from psychoacoustical experiments and from DPOAE measurements commonly show an almost linear growth at low and high (> 100 dB SPL) levels [Kummer *et al.*, 1998, Dorn *et al.*, 2001, Lopez-Poveda and Alves-Pinto, 2008].
The goal of the present study was develop a simple physical model of the cochlea that is able to account for various aspects of physiological data with a single parameter set. The model was required to fulfill the requirements for high sensitivity, finestructure effects and for nonlinear phenomena as described in literature. It is hypothesized that if key aspects of the healthy cochlea are realistically represented by the model, data of other physiological experiments can also be accounted for. It is further hypothesized that such a model can be used to interpret psychophysical data at the level of the cochlea when applied to simple psychoacoustical paradigms. This approach can be used to investigate which aspects of psychoacoustical data can be accounted for on a peripheral level and which effects require processing at higher stages of the auditor pathway. Such a model would be helpful in the investigation of properties of a healthy cochlea as well as the consequences of cochlea damage and the influence on perception.

4.2 Model description

A one-dimensional, active and nonlinear transmission line model of the cochlea operating in the time domain was used. This macromechanical model approach simplifies the cochlea into a system of single elements with the combined mechanical properties of the basilar membrane, the organ of corti and the mass loading of the fluid (cochlea partitions). The model is based on previous work by van den Raadt and Duifhuis [1990], van Hengel *et al.* [1996], Mauermann *et al.* [1999a] and Talmadge *et al.* [1998]. It was shown that this class of nonlinear models can be used to study pure-tone excitation patterns, various aspects of otoacoustic emissions and finestructure effects [e.g. Neely and Kim, 1983, 1986, Talmadge *et al.*, 1998, Mauermann *et al.*, 1999a,b, Moleti *et al.*, 2009].
In order to obtain physiologically plausible excitation patterns that are tall and broad, the nonlinearity and the active process is implemented following the impedance described by Zweig [1991] and in a modified version used in Mauermann *et al.* [1999a]. The model was discretized into 1000 segments, equally spaced along the length of the cochlea. The first and last segment provide boundary conditions for the system: The first segment acts as an implementation of a simple middle ear model to propagate sound into the cochlea and to propagate the activity in the cochlea into the ear canal. The last segment allows to modify the impedance of the helicotrema. The equation of motion for a deflection ξ at a position x along the cochlea is described by the following equation (see Table 4.1 for detailed choice of parameters, points over symbols indicate time derivatives):

$$p(x) = m\ddot{\xi}(x) + d(x,\dot{\xi})\dot{\xi}(x) + s(x)\left[\xi(x) + c(\dot{\xi})\xi(t)|_{t-\tau}\right] \quad (4.1)$$

This second-order differential equation describes the motion of a system consisting of a mass m, a position and velocity dependent damping $d(x,\dot{\xi})$ (damping profile), a position dependent linear stiffness $s(x)$ and a velocity dependent feedback stiffness term $c(\dot{\xi})$ [Mauermann *et al.*, 1999a]. The system is driven by a force caused by the pressure $p(x)$, describing the difference in pressure in the scala tympani and scala media. Velocity dependent damping in combination with the feedback stiffness term model the active behavior and the compressive characteristics of the cochlea. In contrast to previous studies, the nonlinear damping is described by a double sigmoidal (Boltzmann) function on a logarithmic velocity scale with negative values for small velocities and a positive saturation at large velocities. The double sigmoidal damping profile is slightly more complex than a single sigmoidal function, but it turned out that a single sigmoidal damping profile did not fulfil the required properties (see Appendix A). This damping profile can subdivided into four sectors: i) the lower sector with negative damping, ii) the plateau, iii) the middle sector with the slope between plateau and positive saturation and iv) the upper sector with the region of positive saturation (Fig. 4.1). A change of the shape in the different sectors will have effects on the specific characteristics of the model. The function is parameterized to be able to control the different sectors almost independently:

$$\begin{aligned} d(x,\dot{\xi}) = \{ & (1-\gamma)\cdot(\delta_{sat}-\delta_{neg})\cdot \left[1-\frac{1}{1+e^{\frac{\Lambda-\alpha}{\mu_\alpha}}}\right] \\ & +\gamma\cdot(\delta_{sat}-\delta_{neg})\cdot \left[1-\frac{1}{1+e^{\frac{\Lambda-\beta}{\mu_\beta}}}\right] \\ & +\delta_{neg}\}\cdot\sqrt{ms(x)} \end{aligned} \tag{4.2}$$

The function is evaluated for the velocity level in dB given by the variable Λ:

$$\Lambda = 20\cdot\log_{10}\left(\frac{|\dot{\xi}|}{\xi_0}\right), \quad \xi_0 = 10^{-6}\frac{m}{s} \tag{4.3}$$

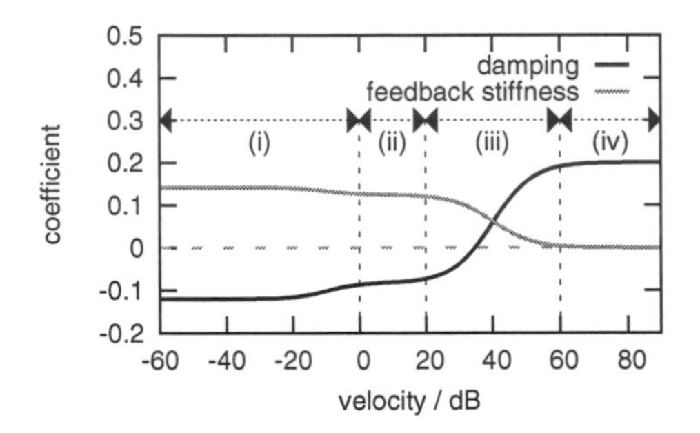

Figure 4.1: Velocity dependent damping profile (solid black line) and feedback stiffness term (solid grey line) as a function of the velocity level. The damping profile can be subdivided into four segments indicated by i-iv. The damping is negative for small velocities (sector i, active part) and increases with increasing velocity (sector iii) with an intermediate plateau (sector ii) to a constant positive damping coefficient (sector iv, passive part). The stiffness term covaries with the damping, approaching zero towards higher velocity levels. While the model is maximal active in sector i, it acts almost linearly in the regime of sector iv.

The velocity dependent part of this function has a value of δ_{neg} for small values and saturates to a value of δ_{sat} for higher levels with an intermediate plateau given by the value γ. Figure 4.1 shows the damping profile (solid black line) and the feedback stiffness term (solid grey line) as a function of the velocity level. To account for changes in Q-factor from base to apex, the saturation value of the damping profile is smoothly changed from base to apex [Talmadge *et al.*, 1998] which is in agreement with psychoacoustical data [Moore, 1978]. The position and the slopes in the middle and the lower sector are given by the parameters α, m_α and β, m_β in equation 4.2, respectively. The stiffness $s(x)$

determines the place-frequency (x-f) map following the shape as proposed by Greenwood [1961]:

$$f_R(x) = \frac{1}{2\pi}\sqrt{\frac{s(x)}{m}} = \mathrm{A} \cdot 10^{-\lambda \cdot x} - \kappa \tag{4.4}$$

The feedback stiffness term covaries with the nonlinear damping term and vanishes at high velocity levels:

$$c(x,\dot{\xi}) = \{(1-\gamma)\cdot(\sigma_{zweig})\cdot \left[\frac{1}{1+e^{\frac{\Lambda-\alpha}{\mu_\alpha}}}\right] + \gamma\cdot(\sigma_{zweig})\cdot \left[\frac{1}{1+e^{\frac{\Lambda-\beta}{\mu_\beta}}}\right]\}$$

For the finestructure simulations, a perturbation in the mechanical properties of the cochlea was introduced in the stiffness term [Zweig and Shera, 1995]:

$$\tilde{s} = s(x)\cdot[1+\epsilon\cdot\mathcal{N}(0,1)] \tag{4.5}$$

The perturbations were Gaussian distributed with zero mean and variance of unity, scaled by the parameter ϵ. The shape of the damping function with its three parts characterize the nonlinear behavior of the model as a function of velocity level. The negative damping in connection with the delayed feedback stiffness at low velocities models the active behavior of the cochlea. This active behavior of the cochlea is a necessary condition to obtain realistic tall and broad excitation patterns. This in turn is, together with the roughness, a prerequisite for finestructure effects [Zweig and Shera, 1995]. An energy source in the cochlea is also necessary to obtain sustained oscillatory behavior in the cochlea without ongoing external excitation. This oscillation is the result of the interplay of multiply reflected waves amplified by the negatively damped oscillators of the cochlea partitions. These oscillations can propagate through the middle ear and can be measured in the ear canal as spontaneous otoacoustic

Table 4.1: Parameters used for the simulations. The parameters were fixed for all simulations.

Parameter	Value	Description
Middle ear		
A_{st}	$3{\cdot}10^{-6}$ m^2	Stapes Area
A_{tm}	$60 \cdot 10^{-6}$ m^2	Tympanic membrane area
ME_Q	0.4	Middle ear quality factor
ME_N	30	Middle ear transformer
Cochlea		
x	$0 \leq x \leq 35 \cdot 10^{-3}$m	Longitudinal coordinate
$\tilde{x}$	$0 \leq \tilde{x} \leq 1$	Normalized length
m	0.375 kg m^{-2}	Membrane mass per unit area
$\tau(x)$	$1.742/f_R(x)$	Delay of feedback stiffness
γ	0.12	Amplitude of plateau
δ_{sat}	$0.2{\cdot}10^{0.52\tilde{x}}$	Damping saturation
δ_{neg}	$-0.12{\cdot}10^{-0.17\tilde{x}}$	Negative damping
α	40	Middle turning point velocity
μ_α	6	Slope at middle turning point
β	-10	Lower turning point velocity
μ_β	5	Slope at lower turning point
A	20832 s^{-1}	x-f map coefficient
λ	60 m^{-1}	x-f map length constant
κ	145.5 s^{-1}	x-f map correction
σ_{zweig}	$0.1416{\cdot}10^{-0.17\tilde{x}}$	Feedback stiffness amplitude
ϵ	0.01	roughness scaling coefficient

emissions (SOAE). The plateau in the damping profile is necessary to limit the amplitude of this sustained oscillations to a plausible level and to limit the amplitude of the excitation patterns for low excitation levels to a range with (almost) constant damping coefficient. This is necessary to obtain a linear, but still active, behavior for low levels as seen in input-output functions of the cochlea [Ruggero *et al.*, 1997, Nelson *et al.*, 2001]. It was argued that the active process has to vanish in the vicinity of stapes and helicotrema in order to obtain stability [Mauermann *et al.*, 1999a]. This is realized by smoothly reducing the negative damping at base and helicotrema to the corresponding saturation values, making the cochlea passive at these points. The slope of the transition between the plateau and the positive saturation of the damping function determines the exponent and the range of the compressive growth of the I/O function.
Fluid coupling is assumed between the single cochlea partitions with the fluid regarded as linear, incompressible and inviscid. The coupled system was solved using Gauss elimination and the differential equations are solved using a 4th-order Runge-Kutta method with a computational frequency of 400 kHz to reduce temporal sampling artifacts.
The middle ear was realized as a simple combination of a mass, stiffness and damping elements with a transformer, leading to a bandpass transfer function peaked at 2 kHz with a quality factor of 0.4. The loading impedance as seen by the cochlea can be adapted to simulate the presence of an earcanal coupler.

4.3 Model evaluation

Before applying the model to psychoacoustical and OAE paradigms, the ability of the model to account for some key aspects of the healthy cochlea was evaluated. At first, the response to pure tones of different levels was tested. In a second step, the

I/O function of the model was derived from the excitation patterns for comparison with experimental data.

4.3.1 Response to pure tone stimulation

In-vivo responses of a healthy cochlea show a sharp peak at low stimulus levels [e.g. Rhode, 1971, Ruggero *et al.*, 1997], reflecting the contribution of the "cochlea amplifier" [Davis, 1983]. The response of the cochlea at the place with maximum excitation at low excitation level (characteristic place, CP) was shown to be almost linear for low excitation levels, to be compressive for intermediate levels and to linearize for high excitation levels. This is in line with data on I/O functions derived from a psychoacoustical forward masking experiment [Oxenham and Plack, 1997]. While physiological data is mostly limited to the evaluation of a single point on the basilar membrane (place-fixed view), the evaluation of the model can also be done for all segments for a fixed frequency and varied levels (frequency-fixed view). The frequency-fixed view help to evaluate the model response of the whole length of the cochlea.
To evaluate the ability of the model to account for these aspects, the model was stimulated with a 1-kHz pure tone with levels ranging from 0 to 120 dB SPL. The stimulus had a length of one second including a 20 ms raised-cosine onset ramp. The maximum velocity of each segment over the whole simulation time was taken for evaluation (velocity excitation pattern). Figure 4.2 shows the excitation pattern of the model for the segment velocity level plotted over the distance from the stapes. At locations towards the base from the CP, the excitation patterns show a linear scaling with increasing excitation level. Near the CP, the patterns are broad and tall for low levels, and the basal slope get shallower towards higher levels. The point of maximum excitation moves towards the base for increased levels. At the apical side of the peak, small wiggles show up due to reflections at the helicotrema. The slopes of the patterns are steeper

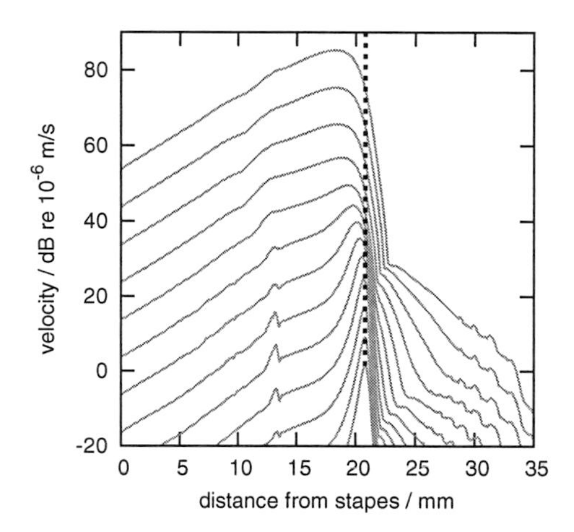

Figure 4.2: Frequency-fixed view of the model response to 1-kHz pure tone stimulation with levels from 0 to 120 dB SPL. Shown is the velocity excitation pattern of all segments along the length of the cochlea. The pattern is plotted as velocity level over the distance of each segment from the stapes. The patterns are broad and tall near the characteristic place and grow with stimulus level. The peak broadens for excitation levels from about 40 to 90 dB SPL and shows linear scaling for higher excitation levels. The point of maximum excitation moves towards the base for increased levels (dashed line). At the apical side of the CP the slopes are steeper than at the basal side. The excitation patterns marked in grey from bottom to top are for stimulus levels of 20, 60 and 90 dB SPL, respectively.

at the apical side of the peak than at the basal side.
In order to compare the level-dependence of the model to physiological, place-fixed measurements, an I/O function was derived from the excitation patterns by replotting the velocity level at a fixed place (CP of the 1-kHz tone at 0 dB SPL) as a function of the excitation level. The resulting I/O function is shown in Fig. 4.3. The response of the model grows almost linearly with a slope of about 0.76 dB/dB for low levels, shows a strongly compressive region for intermediate levels with a slope of 0.26 dB/dB and approaches a linear growth for levels above about 100 dB SPL. The overall compression of the input range is 50 dB. The slope of the growth function for low levels depends on the turning point of the upper sigmoidal function (parameter α). A shift of this turning point towards higher levels lead to a more horizontal plateau and hence in a more linear growth for low levels. A change of the slope of the upper sigmoid m_α changes the rate of compression for intermediate levels. Steeper slopes lead to higher compression. An adjustment of the saturation value d_{sat} could be used to control the velocity range of compressive growth.
This shape of the I/O function shares many characteristics of experimentally derived I/O functions. The lower point where the growth function gets compressive as well as the compression rate is in agreement with physiological data from guinea pigs and chinchilla [Robles and Ruggero, 2001, for a review] and also with data from experimentally derived I/O function for human ears [Nelson *et al.*, 2001, Johannesen and Lopez-Poveda, 2008]. The upper level where the compression decreases is also in agreement with the observation that compression can still be found for levels as high as 90 dB SPL.
There are also physiological data that indicate a different shape of the I/O function. At basal sites of the chinchilla cochlea, the I/O function shows no linear regime at very low levels in some data. This might be supported by that fact that DPOAEs, for which nonlinear effects are apparently necessary, can be mea-

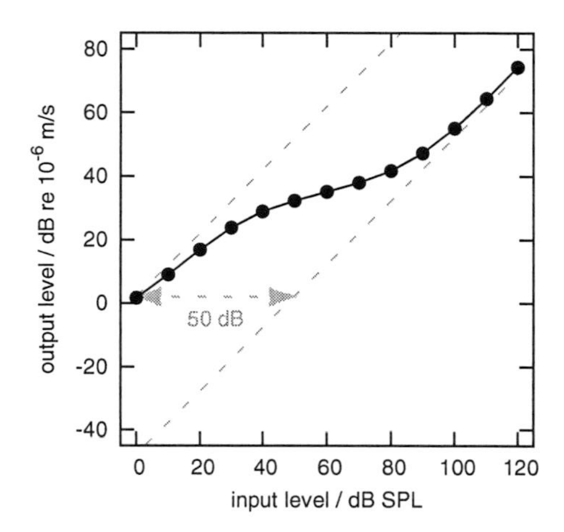

Figure 4.3: Input-output curve at the CP of a 1-kHz pure tone with a level of 0 dB SPL. The grey dashed lines are separated by 50 dB and indicate a linear growth (1 dB/dB). The points indicate the velocity output level of the segment at the CP for the corresponding input level of the pure tone. The curve grows almost linearly with a slope of about 0.76 dB/dB for levels between 0 and 30 dB SPL, shows a compressive region with a slope of about 0.26 dB/dB for intermediate input levels and a linear growth for high input levels. The compressive growth for intermediate levels results in a compression of the dynamic range of about 50 dB.

sured at very low levels [e.g. Kummer *et al.*, 1998]. For the model application part of this study, the shape of I/O function as shown in Fig. 4.3 was chosen as it supports many aspects of I/O functions derived from both, physiological and psychoacoustical experiments. A different shape can be accomplished by variation of the shape of the damping profile.

4.4 Model application

Excitation patterns and I/O-functions found in the model are in agreement with experimental data, indicating that the model reflects key aspects of real cochlea dynamics. Now the model can be applied to other experimental paradigms to assess the validity of the model. At first, the nonlinearity of the model was evaluated in a two-tone suppression paradigm. In a second step, a DPOAE growth function of the model cochlea was simulated. The influence of the roughness was evaluated in the ear canal as stimulus frequency otoacoustic emission (SFOAE) finestructure. The fourth application of the model was to simulate a threshold in quiet with introduced roughness to get threshold finestructure. As a last application, the psychoacoustical paradigm of Heise *et al.* [2009b] was used to investigate the representation of amplitude modulation at the level of the cochlea.
For the evaluation of otoacoustic emissions, the influence of an ear canal coupler was simulated by changing the load impedance of the cochlea to the ear canal.

4.4.1 Response to two-tone stimulation

Physiological as well as psychoacoustical data indicate that the response of the cochlea shows nonlinear aspects when exposed to multi-tone stimuli [e.g. Duifhuis, 1980, Ruggero *et al.*, 1992, Geisler and Nuttall, 1997]. One aspect of nonlinear in-

teraction between different frequency components is the effect of suppression. Suppression was defined as the difference in the excitation due to one tone (suppressee) alone and in the presence of a second tone (suppressor). When both, suppressor and suppressee are pure tones, this effect is often referred to as two-tone suppression. For conditions where the frequency of the suppressor is higher than the frequency of the suppressee, the effect is referred to as high-side suppression and it is called low-side suppression when the suppressor frequency is lower than that of the suppressee. Suppression can occur as a reduction in the overall amplitude (tonic suppression) or only during certain phases of the time course (phasic suppression) [Geisler and Nuttall, 1997]. If the nonlinearity which is embedded in the model reflects realistic aspects of cochlea mechanics, the model should be applicable to an experimental two-tone suppression paradigm.
To evaluate two-tone suppression in the model, the excitation pattern at the CP of the suppressee alone was evaluated for different frequencies and levels of the suppressor. The difference between excitation with and without suppressor was taken as a measure for suppression. Positive differences will be referred to as excitation and a negative differences as suppressionThe suppressee was a 1-kHz pure tone with a level of 40 dB SPL. Frequency and level of the suppressor were varied over six octaves from 0.5 to 2 kHz and from 10 to 120 dB SPL. The stimulus had a length of one second including 20 ms raised-cosine onset ramps. Figure 4.4 shows the simulated suppression for the 1-kHz suppressee. The data is plotted with contour lines, positive values indicating excitation (grey dashed lines) and negative values suppression (solid black lines). For suppressor frequencies lower than that of the suppressee, only excitation was found. For frequencies higher than that of the suppressee, suppression was found for a broad region, while suppression of more than -3 dB was limited to a narrow frequency and level range. The maximum suppression found for this frequency was -10.9 dB for

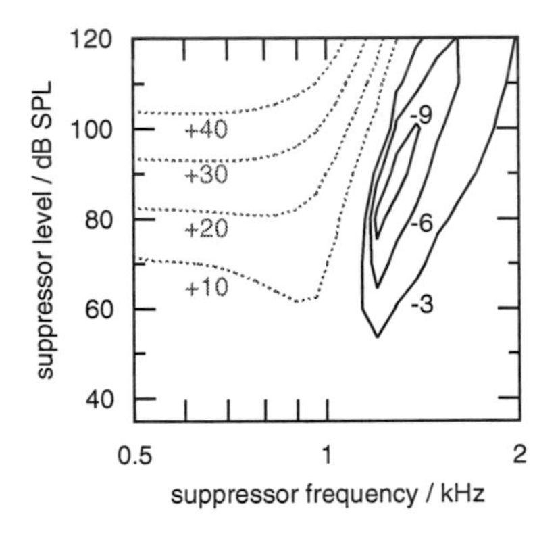

Figure 4.4: Two-tone suppression map for a suppressee frequency of 1 kHz and 40 dB SPL. Excitation is indicated by positive values (dashed grey lines) and suppression as negative values (solid black lines). The maximum high-side suppression was -10.9 dB and low-side suppression was positive for all frequencies and levels of the suppressor.

a suppressor frequency of 1280 Hz and 90 dB SPL. Suppression larger than 3 dB was found in a frequency range of about 1 to 2 kHz for levels between 60 to 120 dB SPL.
For low-side suppression, the excitation of the basliar membrane can show a modulation with the frequency of the suppressor [Ruggero *et al.*, 1992, Geisler and Nuttall, 1997] (phasic suppression). With the excitation pattern as the criterion, suppression is not visible in the model if the peak of the modulated signal is not reduced (tonic suppression). To test for phasic suppression in the model, the temporal dynamics of the segment at the CP of the suppressee was analyzed. The frequency component of the suppressee was extracted in the spectral domain and the level of this component was evaluated. Figure 4.5 shows the amplitude of the suppressee-frequency component with the total excitation at the CP of the suppressee. While the overall excitation (indicated by the excitation pattern in Fig. 4.5) increases almost linearly, the level of the suppressee frequency component decreases with increasing level of the suppressor. This finding is in line with the observation in physiological studies at the base of the guinea pig cochlea [Geisler and Nuttall, 1997]. For levels below 30 dB SPL hardly any influence of the suppressor on the activity of the suppressee was found. Strong suppressive effects were found for levels above about 50 dB SPL. At higher suppressor levels the suppression saturates to a maximum value of about -40 dB at a suppressor level of 100 dB SPL.
These results are in qualitative and quantitative agreement with experimental data [Geisler and Nuttall, 1997] and show that the nonlinearity of the model reflects nonlinear effects across different cochlea segments. This shows that, in contrast to filterbank model, the single segments are not independent of each other, but responses in different frequency regions interact with each other.

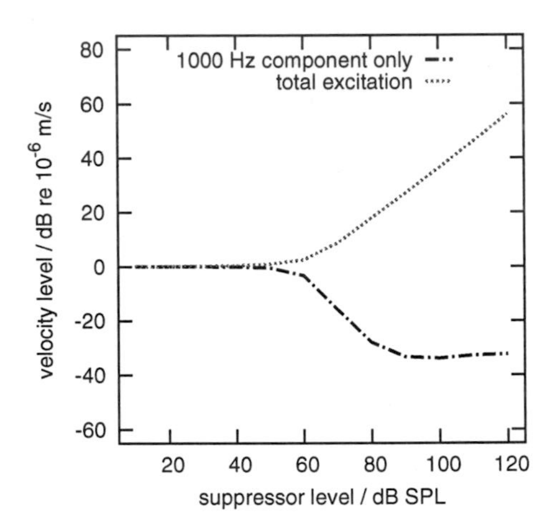

Figure 4.5: Low-side suppression evaluated by separation of the frequency component of the suppressee (solid line). The suppressor had a frequency of 0.5 kHz and the suppressee a frequency of 1 kHz and a level of 40 dB SPL. While the level of the suppressee frequency component decreases, the maximum of the excitation pattern, i.e. the total excitation (dashed line) increases.

4.4.2 DPOAE growth functions

A further consequence of the nonlinearity of the cochlea is the existence of combination tones, which can be measured as distortion product otoacoustic emissions (DPOAE) in the ear canal. It was proposed that the growth of DPOAEs as a function of level of the primary tones (DP-growth function) can be used to characterize the nonlinearity, i.e. to assess the shape of the I/O-function from the DP-growth function [Johannesen and Lopez-Poveda, 2008]. If the I/O function of the model derived from the excitation patterns and the nonlinearity are in line with realistic cochlea mechanics, DP-growth functions should also reflect aspects of experimental data.
To assess the DP-growth function, the model was stimulated with two tones (primaries) with a constant frequency ratio of $f_2/f_1 = 1.2$ and three values of the f_1 primary (2 kHz, 3 kHz, 4 kHz). The level of the primaries was adjusted to maximize the level of the DPOAE, following the scissors paradigm $L_1 = 0.4 \cdot L_2 + 39$ dB proposed by Kummer *et al.* [1998]. The duration of the excitation was one second with 20 ms onset ramps and the pressure in the ear canal with attached coupler was simulated. The level of the $2f_1 - f_2$ component was derived from the spectrum of the last 500 ms. Figure 4.6 shows the growth function of the $2f_1$-f_2 component plotted over the level of the f_2 primary (L_2) for the three different values of the f_1 primary. While there are small differences between the growth functions with the different primary frequencies, the shape of the curve is very similar. For L_2 levels of 20 to about 60 dB SPL, the DP component grows almost linearly from -15 to about 20B SPL and saturates at this level for higher primary levels. The simulation results show that already for low primary levels distortion products arise. The DPOAEs at these low levels are the result of the increasing damping, i.e. the positive slope of the damping profile. The saturation of the DPOAEs results from the constant value of the damping for high velocity levels where the damping

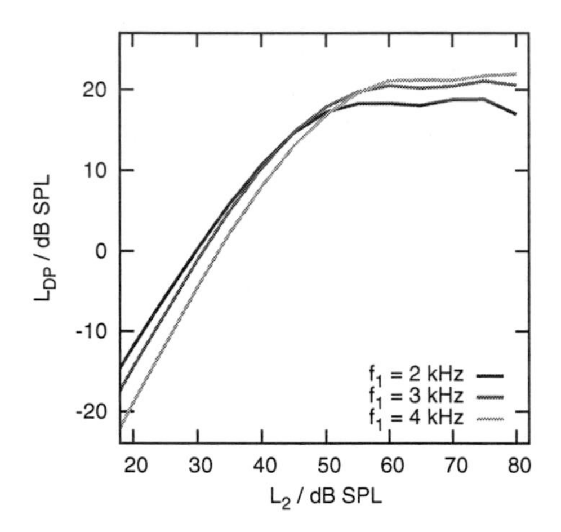

Figure 4.6: Level of the $2f_1 - f_2$ DPOAE distortion component (wave-fixed component only) for different frequencies of f_1 component and a constant frequency ratio $f_2/f_1 = 1.2$ as a function of the level L_2 (DP-growth function). The level of the DPOAE increases compressively with increasing L_2.

profile saturates. In this regime, no further distortion is introduced into the oscillations. The log-like shape of the DP-growth function is in line with data from literature [Mauermann and Kollmeier, 2004, Lopez-Poveda and Alves-Pinto, 2008], but the maximum value of 20 dB SPL is larger than in some experimental data [Mauermann and Kollmeier, 2004].

4.4.3 Finestructure in SFOAE

Since finestructure effects are highly sensitive to properly working active and nonlinear processing in the cochlea, stimulus frequency otoacoustic emissions were simulated to assess the performance of the model at low stimulus levels where the active process is dominant. If the active process properly models the active behavior in a healthy cochlea, the broad and tall excitation patterns should lead to linear reflections and hence to finestructure effects.
A common assumption is that various types of OAEs are the result of coherent reflection of incoming sounds at randomly distributed mechanical inhomogeneities in the peak region of the excitation pattern [e.g. Zweig and Shera, 1995, Talmadge *et al.*, 1998]. The reflected sound wave travels to the stapes where some of the energy is reflected due to the impedance mismatch at the oval window. Multiple reflection will occur that will generate a resonance pattern in the cochlea [Zweig and Shera, 1995]. These resonances enhance the response to some frequencies and reduce the response to others.
Stimulus frequency otoacoustic emissions are emissions evoked by stimulation with a pure tone. This kind of emissions is thought to be a linear reflection of the wave elicited by the stimulus, reflected in the vicinity of the characteristic place at roughness on the cochlea [Zweig and Shera, 1995]. This reflection is most effective in the peak region of the excitation pattern. Experimental data show quasi-periodic fluctuations, i.e. a finestructure in SFOAEs. Next to the multiple reflections within

the cochlea, finestructure of SFOAEs arises in addition due to external interference between the stimulus and the OAE.
To test this mechanism of coherent reflection in the model, SFOAEs were simulated. The stimulus had a length of one second. To separate the stimulus from the SFOAE, the ear canal sound pressure was evaluated for stimulus levels of 20 and 60 dB SPL [Kemp and Chum, 1980, Kalluri and Shera, 2007]. It is assumed that at a level of 60 dB SPL the contribution of the SFOAE is small compared to that of the stimulus, while at 20 dB SPL the relative contribution of the evoked emission is larger. The resulting waveforms in the simulated ear canal were scaled to the same level and subtracted. The last 500 ms of the residual time-pressure signal was multiplied with a Hann window and frequency analyzed. Simulation results are shown in Fig. 4.7. The simulated level of the SFOAE shows peaks and troughs over the simulated frequency range. This difference in SFOAE level indicates the resulting level of a superposition of stimulus and multiple reflections from the peak region within the cochlea, interfering constructively or destructively depending on their relative phase. This finestructure would not be visible without roughness or for a passive cochlea since the reflection sites would be missing and the peaks would not be broad and tall.

4.4.4 Threshold in quiet

One important consequence of the active process in the cochlea is high sensitivity to low levels resulting in a low detection threshold for signals. The performance of the active process was tested by simulating a threshold in quiet. A realistic model of a healthy cochlea should provide a threshold in quiet at a realistic physical level.
To simulate a threshold in quiet, it was assumed that a certain velocity $\dot{\xi}_{crit}$ of the cochlea partition is necessary to trigger a transduction in the inner hair cell that is large enough to detect

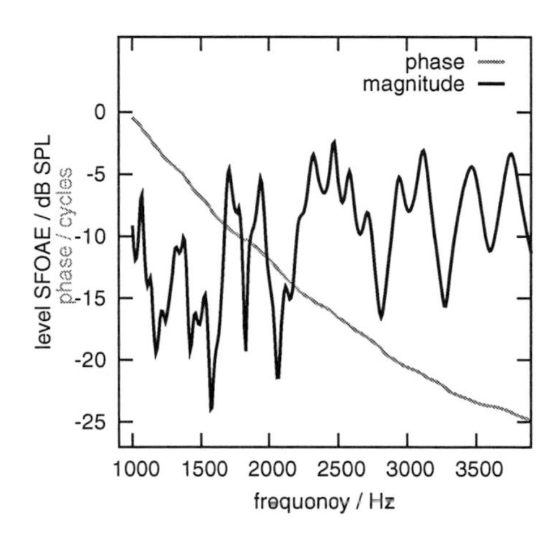

Figure 4.7: Level (black line) and phase (grey line) of the stimulus frequency component (SFOAE) in the simulated ear canal as a function of the stimulus frequency. The SFOAE shows different magnitudes of finestructure and a rapidly rotating phase.

a signal. Velocities below this critical velocity will not trigger transduction and hence the excitation level is not sufficient to detect a signal.

$$P(detect) = \begin{cases} 0 \text{ for } & \dot{\xi} < \dot{\xi}_{crit} \\ \\ 1 \text{ for } & \dot{\xi} \geq \dot{\xi}_{crit} \end{cases} \tag{4.6}$$

For this simplified criterion it was assumed that the inner hair cells (IHC) are (i) not frequency specific, i.e. are identical from base to apex and (ii) that the IHC transduce the velocity of the cochlea partition into a neural signal. This assumption is a simplification to reduce the number of parameters. However, some physiological findings indicate that, e.g., temporal properties of the inner hair cells may change with best frequency [Davis, 2003, Zhou *et al.*, 2005, Liu and Davis, 2007]. A more realistic model may thus include a frequency selective transduction criterion.
Experimental data show that the current into IHC as a function of displacement of the hair bundles follows a sigmoidal function [Russell *et al.*, 1986]. The assumption of a critical velocity is equivalent to the approximation of the sigmoid-shaped transduction function of the inner hair cells by a step-function.
To simulate finestructure effects, roughness was included in the model according to eq. 5. With the introduction of the roughness, a sustained and stable oscillatory activity in the cochlea is introduced. This oscillatory activity can be initiated by a low level broadband excitation (white noise at -50 dB SPL). After the excitation is switched off, the oscillatory activity remains stable in the absence of external stimulation. As an estimate of the critical velocity for the IHC transduction, a level was chosen that was slightly above the steady-state excitation of the model with roughness and without external stimulation. Velocity and displacement in this steady state were also used as initial conditions for the simulations to estimate the threshold in quiet.
Figure 4.8 shows the excitation patterns of the model in the steady state without external stimulation (thick grey line) and

for additional stimulation with a 1-kHz pure tone at 0 dB SPL (solid black line). The excitation pattern for the 1-kHz pure tone with 0 dB SPL without roughness term in the model is shown with a dotted black line. The critical velocity is indicated by the horizontal grey line. The sustained oscillations might also be present in human cochlea as the source of spontaneous otoacoustic emissions (SOAE) measured in the ear canal. The addition of a 1-kHz pure tone to the steady state of the cochlea with roughness hardly changes the excitation pattern at places away from the CP. It is interesting to note that, while the tip of the excitation pattern of the 1-kHz tone without roughness is below the critical velocity, the excitation pattern for the same stimulus with introduced roughness lies above the critical velocity. Thus, the roughness improves the sensitivity of the system for specific frequencies.

To obtain an audiogram the model was stimulated for pure tone with frequencies between 0.9 and 2.5 kHz with a frequency resolution of 1/300th octave. The tones had a duration of one second including a 20-ms raised-cosine onset ramp and the level was varied until the peak of the excitation pattern was within a range of 1 dB to the critical velocity. The initial level for the threshold estimation of each frequency was set to 5 dB above the threshold level of the previous frequency, i.e. the threshold was approached from high levels. The simulated SOAEs were obtained by the frequency analysis of the simulated pressure in the ear canal when the model oscillates in its steady state without external stimulation, i.e. the oscillations underlying the profile in Fig. 4.8. Figure 4.9 shows the simulated audiogram and the SOAE. The simulated threshold (black line) shows a quasi-periodic finestructure at levels around 0 dB SPL in the frequency range from 0.9 Hz to 2.5 kHz. The finestructure shows regions with large (10-14 dB), medium (6-9 dB) and small (2-5 dB) differences between maxima and minima. The frequency components present in the ear canal pressure of the model without external stimulation show levels up to -10 dB

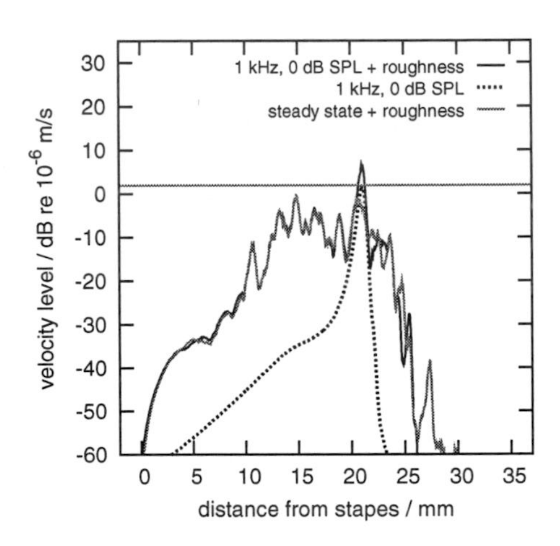

Figure 4.8: Excitation patterns of the model with roughness (solid lines) and without roughness (dashed line). The solid grey line represents the excitation pattern of the model in a steady state after excitation with a broadband stimulus of low intensity. The solid black line indicates the excitation pattern with additional stimulation by a 1-kHz pure tone at 0 dB SPL. The horizontal line represents the critical threshold value assumed in the model for transduction in the inner hair cell. The excitation pattern for a 1-kHz pure tone at a level of 0 dB SPL is shown by the dotted line.

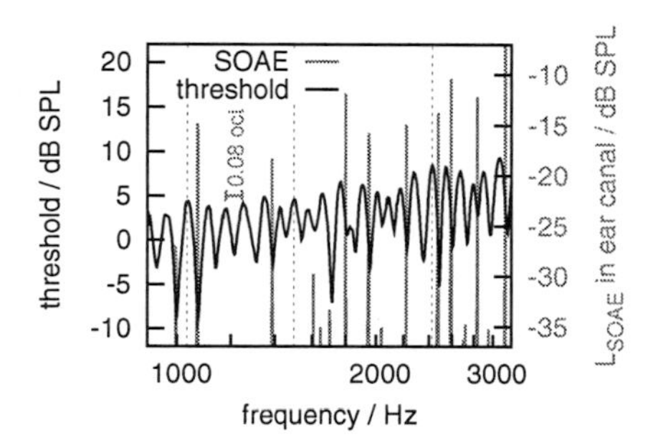

Figure 4.9: Simulated threshold in quiet (black line) and frequency analysis of the simulated ear canal pressure in the steady state of the model (grey lines). The grey horizontal bar indicates the spacing of the simulated finestructure of 0.08 octaves. The simulated threshold in quiet shows quasi-periodic fluctuations with frequency, i.e. a threshold finestructure. The frequency components of the simulated ear canal pressure without external stimulation coincide with finestructure minima. Dashed vertical lines indicate from left to right maxima in regions of large, small and medium finestructure, respectively.

SPL. The frequencies of the SOAE (grey lines) coincide with the frequencies of high sensitivity of the cochlea, i.e. the finestructure minima.
The simulated audiogram shows a realistic shape and magnitude of the threshold in quiet within the simulated frequency range. The periodicity of 0.08 octaves is in line with data found in the literature [Heise *et al.*, 2008, for a review]. The model also accounts for SOAE and the connection to threshold finestructure. These results together with the SFOAE finestructure simulations show that the model reflects a realistic behavior of the active process in the cochlea leading to broad and tall excitation patterns which are a prerequisite for finestructure effects.

4.4.5 Modulation detection threshold and threshold finestructure

It was shown by Heise *et al.* [2009b] that modulation detection thresholds for sinusoidally amplitude modulated tones with low carrier levels depend on the relative position of the spectral components within the threshold finestructure. With roughness, the model shows realistic threshold finestructure in the region of 0.9 Hz to 2.5 kHz. If the differences in modulation detection performance result from the modulation depth at the level of the cochlea, differences in the dynamics of the cochlea partitions in the model with finestructure should be observable. To test this this hypothesis, the model was stimulated with sinusoidally amplitude modulated tones and the modulation depth at the level of the cochlea was analyzed.
The three spectral configurations were the same as used in the psychoacoustical paradigm by Heise *et al.* [2009b,a]: (a) The carrier frequency was at a maximum of the finestructure and the sidebands fell into adjacent minima. the carrier had a level of 15 dB SL (MAX condition). (b) The carrier frequency was at a minimum of the threshold finestructure and the sidebands

at the adjacent maxima. The carrier had the same SPL as in the MAX condition (Min-SPL condition). (c) The carrier- and sideband frequencies were the same as in (b), but the level of the carrier was at the same SL as in the MAX condition (Min-SL condition). In the simulation, the modulation depth at the level of the cochlea was extracted from the frequency analysis of the segment at the CP in each configuration with a fully modulated tone (carrier segment). While the sidebands of the SAM stimuli are symmetrical in level, after cochlear processing the motion of the basilar membrane might show a different spectrum when the dynamics of the segment is analyzed in the frequency domain. For this reason, the average level of the sidebands relative to that of the carrier frequency was used to evaluate the resulting modulation depth. The initial modulation depth in each condition was 0 dB and was adapted until the modulation depth at the level of the cochlea was within a range of 0.5 dB of a critical modulation depth. The critical modulation depth was set to -18 dB. This value was chosen to adjust the model to the MAX condition for large finestructure region. The center frequencies of the stimuli are indicated by the vertical dashed lines in Fig. 4.9. The frequencies of the carriers were 1035 Hz, 2425 Hz and 1500 Hz for large (>10 dB), medium (5-10 dB) and small (<5 dB) finestructure, respectively.

Figure 4.10 shows experimental data from [Heise *et al.*, 2009b] (left panels) and simulated modulation detection thresholds (right panels) for the three conditions. Modulation depth on the level of the cochlea was simulated in regions of large, medium and small threshold finestructure (from top to bottom). The difference between maximum and adjacent minimum is indicated by ΔL. Experimental data shows, that modulation detection performance is best in the MAX condition. Performance drops compared to the MAX condition in MinSPL condition and thresholds are highest in the MinSL condition. The simulations are in good agreement with the data. The stimulus modulation depth necessary to obtain a modulation depth of the dynamics

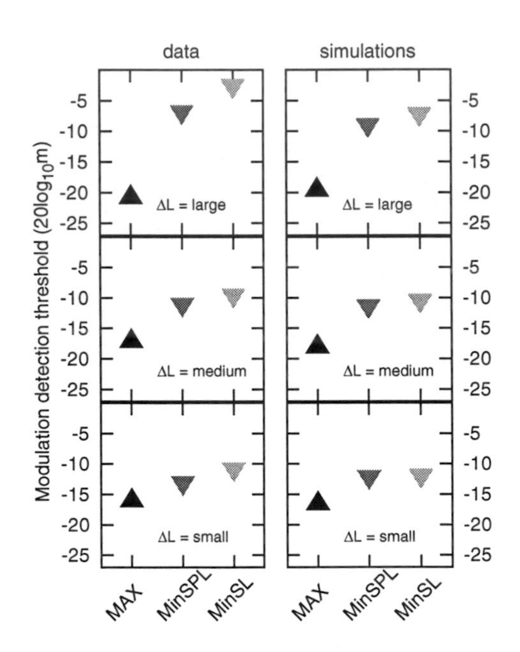

Figure 4.10: Modulation detection thresholds in dB modulation depth. Shown is individual data from Heise *et al.* [2009b] (left panels) and simulated data (left panels). The upper column shows results for a region with large threshold finestructure, the middle panel for low and the lower panel for small threshold finestructure. The magnitude of finestructure (large, medium, small) is indicated by Δ in each panel. The upwards pointing triangles are the results for the MAX condition, the dark grey triangles for the MinSPL condition and the light grey triangles for the MinSL condition

of the carrier segment of -18 dB is smallest for the MAX condition and largest for the MinSL condition. Also in line with the data is the larger difference between MAX and MinSPL than for the MinSPL and the MinSL conditions. In line with the data, the influence of threshold finestructure differs between the three conditions. For the MAX condition, the modulation depth at threshold decreased as the magnitude of finestructure increased (about -15dB for small and -20 dB for large finestructure). On the other hand, for the MinSPL and MinSL conditions, the modulation depth at threshold increased when the magnitude of the finestructure increased. This indicates that finestructure directly influences the temporal dynamics and hence the representation of amplitude modulation on the basilar membrane.

4.5 Discussion

4.5.1 The role of the damping profile in the nonlinear characteristics of the model

The double sigmoid shape of the damping function proved to be useful for modeling physiological data and psychoacoustical data near threshold. In the current model approach, the damping profile with the delayed feedback stiffness is the model of both, the active process and the nonlinearity of the cochlea. A widely accepted view of the active process in the cochlea is that the energy is provided through the motility of the outer haircells (OHC). This energy source was implemented as negative damping for small velocity levels. It is reasonable to assume that this active process is most efficient at low levels and saturates towards higher levels. The transition where the damping changes as a function of the velocity from negative values to positive values determines the nonlinear behavior of the model, defining

the slope and the range of the compression in the I/O function[2]. The negative damping in connection with the roughness lead to self-sustained oscillations of the model. The amplitude of these self-sustained oscillations play a role in the simulated finestructure effects and are the reason for the SOAE found in the simulated ear canal. If the oscillations were transient in nature, i.e., the activity would decay as a function of time, no SOAE could be found in the model. The level of the self-sustained oscillations play an important role not only for the level of the simulated SOAE, but also for the simulated threshold in quiet since the activity due to external excitation must significantly differ from the activity in the absence of external stimulation in order to estimate a threshold in quiet with the simple approach used here.
The balance between sustained oscillations and the nonlinear regime of the model can be controlled via the plateau in the damping profile. This plateau introduced by the second sigmoidal function limits the amplitudes of the sustained oscillations and at the same time linearizes the growth of the excitation patterns at low to intermediate levels, matching the suggestion of a linear-compressive-linear shape of the I/O function. The physiological correlate of this nonlinearity can either be in a more complex dynamics of the organ of corti than a simple strictly monotonic saturation or in two separate mechanisms in the cochlea which are effectively combined in the double-sigmoidal damping profile.

---

[2] One may argue that, for the sake of even a smaller number of parameters, the damping profile could be modeled by a single sigmoidal function rather than by two sigmoid functions as in the present study. This, however, has an influence on the shape of the I/O function and hence on the region and level of two-tone suppression and the DP-growth function and also on the simulated threshold in quiet (see section A) in the appendix.

4.5.2 Evaluation of the nonlinearity

Effects of the nonlinearity in the model was evaluated with the paradigms of two-tone suppression and DPOAE growth function. Both, the simulations of two-tone suppression and DPOAE-growth functions show that the nonlinearity in the cochlea can be combined into a velocity dependent damping function.
Both, low- and high-side suppression was found in the model when the dynamics of the cochlea at the CP of the suppressee was analyzed. From the psychoacoustical point of view, the simulations of the high-side suppression are in line with the assumption that the total excitation of the basliar membrane is attenuated by the presence of the suppressor (tonic suppression). This reduced excitation can be interpreted as an decreased sensitivity because an increase in level of a tone at the frequency of the suppressee is necessary to obtain the same excitation as in the absence of the suppressor. With frequency independent inner hair cell sensitivity, this explains the decreased masking effect of the suppressor on the suppressee in the psychoacoustical pulsation threshold paradigm used to evaluate two-tone suppression in humans [Houtgast, 1972, Duifhuis, 1980]. For low-side suppression, no decrease in the total activity could be found. The presence of the suppressor modulated the dynamics of the segment at the CP of the suppressee such that only a reduction of the velocity component oscillating at the frequency of the suppressee could be found. Under the simplified assumption that temporal aspects are not evaluated by the auditory system for the task of this pure tone detection, the model can not explain low-side two-tone suppression and the psychoacoustical pulsation threshold paradigm is only sensitive to tonic suppression. If however the dynamics of the segment at the CP of the suppressee is analyzed by some kind of frequency selective process at higher stages, then two-tone suppression is covered by the model.

The nonlinearity of the model produced combination tones in the simulated ear canal when stimulated with two-tone stimuli. The level of the $2f_1$-f_2 component of the simulated DPOAEs showed a log-like growth with the level of the primaries. The growth rate and the maximum level of the $2f_1$-f_2 DPOAE component found in the model were higher than found by Mauermann and Kollmeier [2004], but well within the range of individual data from Johannesen and Lopez-Poveda [2008]. The magnitude of the DPOAE depends on the region of overlap of the excitation patterns of the primaries and on the slope of the damping profile. The region of overlap is determined by the width of the excitation patterns of the primaries. Hence, the level of the DPOAE can be adjusted in the model by changing the damping and the slope in sectors ii) and iii) of the damping profile (see Fig. 4.1). Lower positive damping values lead to narrower excitation patterns and steeper slopes lead to higher DP levels and to higher compression in the I/O function. While the slope of the DP-growth function is clearly determined by the damping profile, the absolute value of the OAEs in the model can be influenced by the implementation of the middle ear.

4.5.3 Reflection due to cochlea roughness

The simulation results for the finestructure simulations show that the coherent reflection mechanism contributing to finestructure effects is present in the model. The introduction of roughness leads to complex interactions of activity within the cochlea which are underlying the finestructure effects in the simulated data. The SFOAEs derived from the level in the ear canal show finestructure accompanied with a rapidly rotating phase as observed in experimental data Kalluri and Shera [2007]. The contribution of the transmission of the sound waves through the linear middle ear does not contribute to the level in this relative measure. Hence, the level of the SFOAE is mainly determined by the multiple reflection in the inner ear, which depends strongly

on the amplification at the CP. A necessary condition to get SFOAEs is the reflection of the wave induced by the stimulus at irregularities along the cochlea. Since the magnitude of the reflected wave is dominated by the contribution of the peak of the excitation pattern, the magnitude of the SFOAE is related to the amplification near the CP of the stimulus frequency. The magnitude of the SFOAEs derived from the model response is in line with experimental data. This result supports the strength of the active process in the simulated region.
The same is true for threshold finestructure. The introduction of roughness leads to sustained oscillations in the cochlea. These oscillations are the result of multiply reflected and amplified waves in the cochlea which lead to the SOAE found in the simulated ear canal. The sustained oscillations also lead to differences in sensitivity at different places on the basilar membrane. Removal of the roughness also removes the finestructure effects in the model.

4.5.4 Application to psychoacoustics

Additional support for the active process in the model is given by the realistic values of the simulated threshold in quiet. It is clear that the threshold is connected to the assumption that the inner hair cells are velocity detectors. The definition that some critical velocity must be crossed to detect a tone is certainly a very simplified view. It assumes a binary threshold without any frequency selectivity rather than a psychometric function as underlying more elaborate concepts of signal detection [e.g. Dau *et al.*, 1997]. The simple criterion used here can be modified in order to extend the frequency range over which a threshold in quiet is evaluated. One limiting factor of this approach is that the maximum amplitudes over time of the sustained oscillations are used to simulate a threshold in quiet. Temporal fluctuations in the amplitude by interaction of multiple sustained oscillations, i.e. temporally lower amplitudes of the sustained oscillations are

not resolved by this criterion. Such temporally lower activity in the cochlea might influence the threshold in quiet. Another consequence of this fixed criterion is that SOAEs in the model can never be higher in level than the threshold in quiet for this particular frequency since the excitation pattern of the sustained oscillations would always be higher than that of the external stimulus.
Threshold finestructure is the result of standing waves within the cochlea model which are building up when roughness is introduced in the place-frequency map. It is interesting to note that this results in an increased sensitivity of the system in some regions, i.e., a lower level is necessary to accomplish a certain velocity for this particular region compared to the situation where no roughness is present (see also Fig. B.12). The introduced roughness also leads to sustained oscillations in the cochlea which were evaluated in the ear canal as SOAE. In agreement with experimental data [Zwicker and Schloth, 1984], the frequencies of the SOAE and the minima of threshold finestructure coincide. A decrease in the activity of the active process in the cochlea reduces the magnitude of the SOAEs and also the magnitude of threshold finestructure. Such a decrease also reduces the energy supplied to the travelling wave, leading to a reduction in the velocity level at the peak of the excitation pattern. This supports the hypothesis that threshold finestructure can be used as a sensitive tool to detect early damage of the cochlea amplifier [Heise *et al.*, 2008].

It was hypothesized that the differences in modulation detection performance of sinusoidally amplitude modulated tones can be explained by the representation of the stimulus modulation at the level of the cochlea. The assumptions made for this simulation were similar to those for the simulations of the threshold in quiet. The model simulations shows a realistic threshold including finestructure. With this model applied to the experimental paradigm of Heise *et al.* [2009b], realistic modulation detection thresholds could be simulated. Hence, the realistic threshold finestructure leads to differences in the representation of amplitude modulated tones at the level of the cochlea. This shows that such a model can be used for a detailed investigation of the influence of threshold finestructure on psychoacoustic performance near threshold.

4.6 Summary & Conclusions

A one-dimensional nonlinear and active model of the cochlea with a double sigmoidal damping profile was developed and evaluated. The developed double sigmoidal damping profile serves as the nonlinearity and the active element in the model and accounts for various aspects of the healthy cochlea. The parameterization of the damping profile allows to control the relevant nonlinear and active characteristics of the model. In contrast to filter-based models of the cochlea, the model of the present study combines the ability to account for multiple aspects of preprocessing of sounds within the cochlea as well as the ability to simulate data from physiological experiments and otoacoustic emissions[3]. With a fixed parameter set, the model accounts for 1) an I/O function as derived from psychoacoustical data, 2) high- and low-side two-tone suppression, 3) distortion products of otoacoustic emissions (DPOAE), 4) finestructure effects in stimulus frequency otoacoustic emissions (SFOAE), 5) threshold finestructure including spontaneous otoacoustic emissions (SOAE) in finestructure minima and 6) the performance of sinusoidal amplitude modulation detection near threshold of human listeners.

[3]In the present study, mainly effects of amplitude were investigated. More detailed quantitative investigation of OAEs can be found elsewhere [e.g. Shera, 2003, Kalluri and Shera, 2007]

It could be shown that a model with a limited parameter set can account for a variety of physiological data and psychoacoustical data simultaneously. This strongly supports the hypothesis that some aspects of psychoacoustical performance can already be explained by peripheral preprocessing. Thus, a realistic description of the periphery can help to disentangle peripheral processing at the level of the cochlea and neural processing leading to a more detailed understanding of the processing stages following the cochlea. In the future, such a model may be used to investigate single subject data by individual adjustment of the model parameters and to investigate the influence of cochlea damage on various aspects of the dynamics of the cochlea and the consequences for psychoacoustical performance.

Acknowledgements

This work was supported by the Deutsche Forschungsgemeinschaft (Grant Nos. SFB/TRR31, KO942/18-1&2). The first author was supported by the Heinz-Neumüller Stiftung. The authors would like to thank H. Duifhuis for many fruitful discussions and for providing the source code of the modeling framework.

A Complexity of the damping profile

If a comparable compressive slope of the I/O function is desired with a single-sigmoid damping profile, the damping profile must have the same slope as the double-sigmoid profile in sector iii). In order to obtain a similar compressive range of 50 dB, the range between the linear part of the damping profile for low levels, where the damping is negative, and the linear part for high levels, where the damping function saturates, must also be similar to that of the double-sigmoid damping profile. The negative damping feeds energy into the system until the am-

plitude is balanced by the delayed feedback stiffness. A single sigmoidal function with the negative damping values as suggested by Zweig (1991) lead to large amplitudes of the sustained oscillations with introduced roughness. In order to keep the amplitude of the self-sustained oscillations, and hence the threshold in quiet, at a comparable level for the single-sigmoid profile than for the double-sigmoid profile, the transition point of the damping profile must be shifted towards lower velocity levels (see Fig. A.11). This shift has consequences on the simulated two-tone suppression and the DPOAE growth function. With a shift of the damping profile towards lower velocity levels, the damping saturates at lower input levels and hence the system linearizes at lower input levels, leading to a compressive-linear rather than a linear-compressive-linear I/O function (panel A in Fig. A.11). This would shift the regions of two-tone suppression towards lower levels. The DP-growth function would show larger values for lower levels L_2 and would saturate at lower levels of the primaries. A single sigmoidal function with a transition at higher levels (panel B in Fig. A.11) leads to a linear-compressive-linear I/O function, but the amplitude of the sustained oscillations increase by about 20 dB. This would shift the simulated threshold in quiet to unrealistically high levels. The double sigmoidal damping profile (panel C in Fig. A.11) leads to smaller sustained oscillations with the desired shape of the I/O function.
In order to obtain a monotonically increasing level of the DPOAEs, the model needs to show nonlinearity at higher input levels. This can be obtained by changing the saturation in section iv) of the damping profile to a non-constant behavior. Since not much data exist for DPOAE at high primary levels, in the present study it was assumed that the cochlea linearizes, i.e. the cochlea amplifier saturates at high input levels.

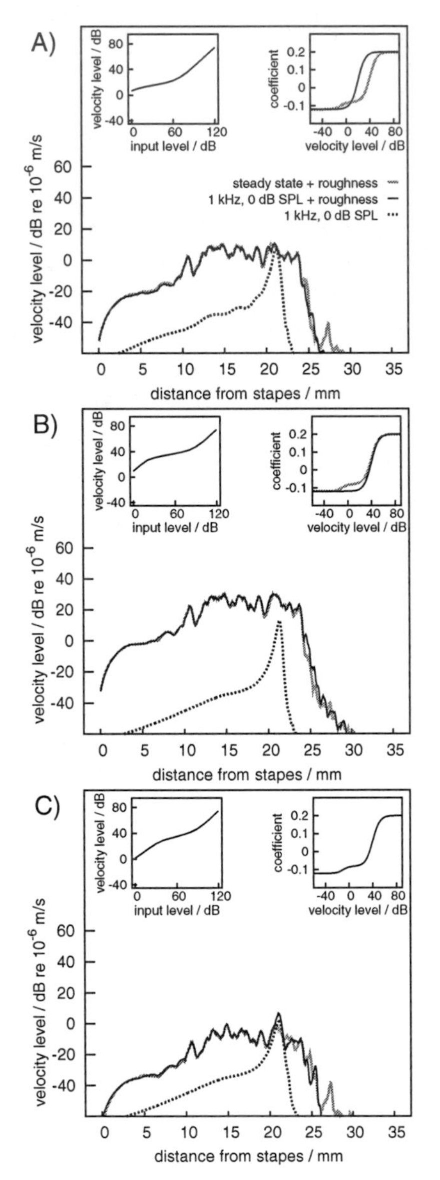

Figure A.11: Excitation patterns of the model for two different single sigmoidal (A,B) damping profiles and the double sigmoidal profile used in the present study (C). The left inset shows the corresponding input-output function and the right inset the damping profile (black) compared to the double sigmoidal damping profile used in the present study (grey).

B Variation of the roughness

The magnitude of the roughness scaling coefficient ϵ in equation 4.5 was adapted from Mauermann *et al.* [1999a]. It was found that the chosen value of 1% yields a plausible magnitude in the simulated threshold in quiet. For smaller values of ϵ, the finestructure vanishes until no finestructure can be found for $\epsilon = 0$. The simulated threshold finestructure for values of $\epsilon \in \{1\%, 0.5\%, 0.1\%, 0\%\}$ is shown in Fig. B.12. For values larger than 1%, the simulated excitation patterns become more noise like and show a larger amplitude. It is possible that for different damping profiles with different values of the negative damping, this value might differ from the value chosen in the present study. Since it is not clear what the physiological correlate of the roughness is, the magnitude of this parameters has to be chosen to match data on finestructure effects.
The roughness was drawn from a normal distribution. Another random seed obviously results in a different perturbation of the place-frequency map, leading to a different finestructure pattern in the threshold in quiet. Hence, in order to simulate finestructure in different cochleae, each newly drawn roughness might be identified with a cochlea of a different subject.

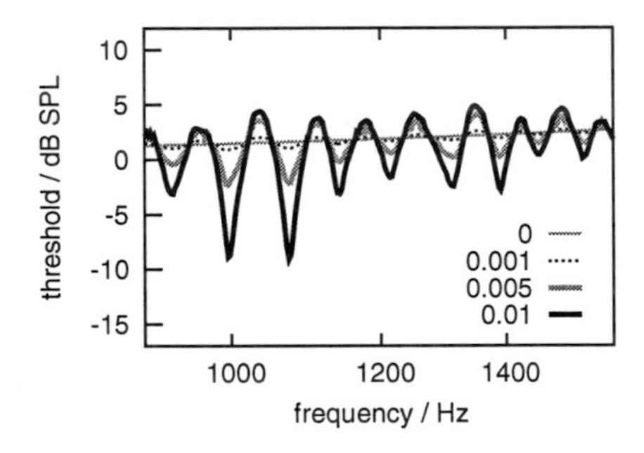

Figure B.12: Simulated threshold in quiet for different values of the roughness scaling coefficient (parameter ϵ in Tab. 4.1). The solid black line indicates the same finestructure as shown in Fig. 4.9. For no perturbation, the threshold is smooth and the magnitude of the finestructure increases with increasing amplitude of the roughness.

5 Summary and concluding remarks

The aim of this thesis was to investigate the role of combined across-frequency and binaural processing in auditory object formation and the influence of cochlear preprocessing on psychoacoustical effects. To quantify the processing of across-frequency and binaural cues, the ability to detect masked signals was investigated. Psychoacoustical experiments with human listeners were performed to assess the efficiency of the auditory system to use both, comodulation and interaural phase disparities separately and in combination. A functional model based on signal processing techniques with a minimal set of parameters was used to investigate the hypothesis derived from the experimental data. As a basis to refine the functional model approach by a detailed analysis of the contribution of cochlear processing, a physical cochlea model was developed which accounts for various physiological and psychoacoustical effects with a single parameter set.

In chapter 2 an experimental paradigm based on a flanking band paradigm commonly used to investigate CMR, was developed to investigate the combination of CMR and BMLD. The results showed that the masked threshold of the tone was lower if either comodulation or an interaural phase difference was present compared to a condition where the cues were absent. In conditions where a combination of comodulation and interaural phase difference was present, the overall benefit was the sum of the benefits due to comodulation and the interaural phase difference separately. A simplified model with a serial alignment of across-frequency envelope and interaural finestructure processing was developed that was able to reproduce the dataset. In chapter 3 it was investigated, whether the conclusion based on results described in chapter 2 could be generalized over a broad frequency range. The results show that the additivity of CMR and BMLD can also be found for center frequencies of 200 Hz and 3000 Hz. The additivity of CMR and BMLD also holds for a transposed stimulus. With this data in the context of the data published by Hall *et al.* [1988], Cohen and Schubert [1991], van de Par and Kohlrausch [1999] and Hall *et al.* [2006] it can now be hypothesized that across-channel CMR is additive to BMLD but that within-channel contributions to CMR are reduced in dichotic listening conditions. The model simulations and the data show that across-frequency envelope information and interaural finestructure information can be used independently. Hence, across-frequency information and spatial information are important cues for the auditory system to form auditory objects which are used in an optimal manner. The experimental data and the model simulations did show that a serial processing of the cues is possible, but do not allow a clear conclusion about the order in which the processing stages are aligned. The developed experimental paradigm may be used as a basis for further experiments with methods of electrophysiology and imaging techniques to give an answer to that question.

In chapter 4, a realistic model of the cochlea was developed to investigate contributions of peripheral processing effects on psychoacoustical data. The simulation results of chapter 4 show that a realistic model of the cochlea with a limited parameter set can be used to account for several important aspects of auditory preprocessing at the level of the cochlea. It was shown that such a model with a simple decision stage can be applied to simulate psychoacoustical data on modulation detection performance near threshold. With such a realistic model of the cochlea that accounts for important aspects of peripheral preprocessing, a more detailed representation of the preprocessing is available that also includes nonlinear level effects and interaction of different frequency components. It is hypothesized that this model can be used to evaluate the contribution of peripheral and more central processing stages to psychoacoustical effects. This provides more detailed insight into the performance of the auditory system to process comodulation and interaural disparities to form auditory objects. The increase in computational power encourages the attempt to include a realistic model of the cochlea into auditory models.

The present thesis showed that comodulation as a spectro-temporal property and the interaural phase resulting from the spatial location of a sound source are important cues for auditory object formation. The intact auditory system can process these cues independently. The neural processing is supported by the nonlinear preprocessing of the active cochlea which already contributes to psychoacoustical effects. This combination of mechanical and neural processing provides humans with the outstanding ability to transform highly complex sound fields into accessible auditory scenes.

Bibliography

Beutelmann, R. and Brand, T. (**2006**). "Prediction of speech intelligibility in spatial noise and reverberation for normal-hearing and hearing-impaired listeners", J. Acoust. Soc. Am. **120**, 331–342.

Breebaart, J., van de Par, S., and Kohlrausch, A. (**2001**). "Binaural processing model based on contralateral inhibition. I. Model structure", J. Acoust. Soc. Am. **110**, 1074–1088.

Buschermohle, M., Verhey, J. L., Feudel, U., and Freund, J. A. (**2007**). "The role of the auditory periphery in comodulation detection difference and comodulation masking release", Biological Cybernetics **97**, 397–411.

Buss, E., Hall, J. W., and Grose, J. H. (**2007**). "Individual differences in the masking level difference with a narrowband masker at 500 or 2000 Hz", J. Acoust. Soc. Am. **121**, 411–419.

Buus, S. (**1985**). "Release from masking caused by envelope fluctuations", J. Acoust. Soc. Am. **78**, 1958–1965.

Cherry, E. C. (**1953**). "Some experiments on the recognition of speech, with one and with 2 ears", J. Acoust. Soc. Am. **25**, 975–979.

Cohen, M. F. (**1982**). "Detection threshold microstructure and its effect on temporal integration data", J. Acoust. Soc. Am. **71**, 405–409.

Cohen, M. F. (**1991**). "Comodulation masking release over a 3 octave range", J. Acoust. Soc. Am. **90**, 1381–1384.

Cohen, M. F. and Schubert, E. D. (**1991**). "Comodulation masking release and the masking-level difference", J. Acoust. Soc. Am. **89**, 3007–3008.

Cooper, N. P. and Rhode, W. S. (**1997**). "Mechanical responses to two-tone distortion products in the apical and basal turns of the mammalian cochlea", Journal of Neurophysiology **78**, 261–270.

Darrow, K. N., Maison, S. F., and Liberman, M. C. (**2006**). "Cochlear efferent feedback balances interaural sensitivity", Nat. Neurosc. **9**, 1474–1476.

Dau, T., Ewert, S., and Oxenham, A. J. (**2009**). "Auditory stream formation affects comodulation masking release retroactively", J. Acoust. Soc. Am. **125**, 2182–2188.

Dau, T., Kollmeier, B., and Kohlrausch, A. (**1997**). "Modeling auditory processing of amplitude modulation .1. Detection and masking with narrow-band carriers", J. Acoust. Soc. Am. **102**, 2892–2905.

Davis, H. (**1983**). "An active process in cochlear mechanics", Hear. Res. **9**, 79–90.

Davis, R. L. (**2003**). "Gradients of neurotrophins, ion channels, and tuning in the cochlea", Neuroscientist **9**, 311–316.

deBoer, E. (**1983**). "No sharpening - a challenge for cochlear mechanics", J. Acoust. Soc. Am. **73**, 567–573.

deBoer, E. (**1995**). "The inverse problem solved for a 3-dimensional model of the cochlea .1. Analysis", J. Acoust. Soc. Am. **98**, 896–903.

Dorn, P. A., Konrad-Martin, D., Neely, S. T., Keefe, D. H., Cyr, E., and Gorga, M. P. (**2001**). "Distortion product otoacoustic emission input/output functions in normal-hearing and hearing-impaired human ears.", J. Acoust. Soc. Am. **110**, 3119–3131.

Duifhuis, H. (**1980**). "Level effects in psychophysical two-tone suppression", J. Acoust. Soc. Am. **67**, 914–927.

Durlach, N. I. (**1963**). "Equalization and cancellation theory of binaural masking-level differences", J. Acoust. Soc. Am. **35**, 1206–1218.

Eddins, D. A. (**2001**). "Monaural masking release in random-phase and low-noise noise", J. Acoust. Soc. Am. **109**, 1538–1549.

Eddins, D. A. and Wright, B. A. (**1994**). "Comodulation masking release for single and multiple rates of envelope fluctuation", J. Acoust. Soc. Am. **96**, 3432–3442.

Elliott, E. (**1958**). "Ripple effect in the audiogram", Nature **181**, 1076–1076.

Elliott, S. J., Ku, E. M., and Lineton, B. (**2007**). "A state space model for cochlear mechanics", J. Acoust. Soc. Am. **122**, 2759–2771.

Epp, B. and Verhey, J. L. (**2009**). "Superposition of masking releases", J. Comput. Neurosc. **26**, 393–407.

Ernst, S. M. A. and Verhey, J. L. (**2006**). "Role of suppression and retro-cochlear processes in comodulation masking release", J. Acoust. Soc. Am. **120**, 3843–3852.

Ernst, S. M. A. and Verhey, J. L. (**2008**). "Peripheral and central aspects of auditory across-frequency processing", Brain Res. **1220**, 246–255.

Ewert, S. D. and Dau, T. (**2000**). "Characterizing frequency selectivity for envelope fluctuations", J. Acoust. Soc. Am. **108**, 1181–1196.

Fletcher, H. (**1940**). "Auditory Patterns", Rev. Mod. Phys. **12**, 47–61.

Geisler, C. D. and Nuttall, A. L. (**1997**). "Two-tone suppression of basilar membrane vibrations in the base of the guinea pig cochlea using "low-side" suppressors", J. Acoust. Soc. Am. **102**, 430–440.

Gold, T. (**1948**). "Hearing. 2. The physical basis of the action of the cochlea", Proceedings of the Royal Society of London Series B-Biological sciences **135**, 492–498.

Green, D. M. (**1992**). "On the similarity of 2 theories of comodulation masking release", J. Acoust. Soc. Am. **91**, 1769–1769.

Green, D. M. and Swets, J. A. (**1988**). *Signal Detection Theory and Psychophysics* (Wiley, New York, NY, 1966 (Reprinted in 1988, Los Altos, CA, Peninsula Publishers)).

Greenwood, D. (**1961**). "Critical bandwidth and frequency coordinates of basilar membrane", J. Acoust. Soc. Am. **33**, 1344–1356.

Griffiths, T. D. and Warren, J. D. (**2004**). "What is an auditory object?", Nat Rev Neurosci **5**, 887–892.

Hall, J. W., Buss, E., and Grose, J. H. (**2006**). "Binaural comodulation masking release: Effects of masker interaural correlation", J. Acoust. Soc. Am. **120**, 3878–3888.

Hall, J. W., Cokely, J. A., and Grose, J. H. (**1988**). "Combined monaural and binaural masking release", J. Acoust. Soc. Am. **83**, 1839–1845.

Hall, J. W. and Fernandes, M. A. (**1983**). "The effect of random intensity fluctuation on monaural and binaural detection", J. Acoust. Soc. Am. **74**, 1200–1203.

Hall, J. W., Grose, J. H., and Haggard, M. P. (**1990**). "Effects of flanking band proximity, number, and modulation pattern on comodulation masking release", J. Acoust. Soc. Am. **87**, 269–283.

Hall, J. W., Haggard, M. P., and Fernandes, M. A. (**1984**). "Detection in noise by spectro-temporal pattern-analysis", J. Acoust. Soc. Am. **76**, 50–56.

Hatch, D. R., Arne, B. C., and Hall, J. W. (**1995**). "Comodulation Masking Release (CMR) - Effects of gating as a function of number of flanking bands and masker bandwidth", J. Acoust. Soc. Am. **97**, 3768–3774.

Heise, S. J., Mauermann, M., and Verhey, J. L. (**2009**a). "Investigating possible mechanisms behind the effect of threshold fine structure on amplitude modulation perception", J. Acoust. Soc. Am. **126**, 2490–2500.

Heise, S. J., Mauermann, M., and Verhey, J. L. (**2009**b). "Threshold fine structure affects amplitude modulation perception", J. Acoust. Soc. Am. **125**, EL33–EL38.

Heise, S. J., Verhey, J. L., and Mauermann, M. (**2008**). "Automatic screening and detection of threshold fine structure.", Int J Audiol **47**, 520–532.

Helmholtz, H. (**1863**). *Die Lehre von den Tone als physiologische Grundlage für die Theorie der Musik* (Verlag Friedrich Vieweg, Braunschweig).

Hirsh, I. J. (**1948**). "The influence of interaural phase on interaural summation and inhibition", J. Acoust. Soc. Am. **20**, 536–544.

Hohmann, V. (**2002**). "Frequency analysis and synthesis using a Gammatone filterbank", Acta Acust./Acustica **88**, 433–442.

Horst, J. W., Wit, H. P., and Albers, F. W. J. (**2003**). "Quantification of audiogram fine-structure as a function of hearing threshold", Hear. Res. **176**, 105–112.

Houtgast, T. (**1972**). "Psychophysical evidence for lateral inhibition in hearing.", J. Acoust. Soc. Am. **51**, 1885–1894.

Hudspeth, A. J. (**2008**). "Making an effort to listen: Mechanical amplification in the ear", Neuron **59**, 530–545.

Hudspeth, A. J. (**2009**). "Making an effort to listen: Mechanical amplification by myosin molecules and ion channels in hair cells of the inner ear", Journal of Physiological Sciences **59**, 17–17.

Ingham, N. J., Bleeck, S., and Winter, I. M. (**2006**). "Contralateral inhibitory and excitatory frequency response maps in the mammalian cochlear nucleus", Eur. J. Neurosci. **24**, 2515–2529.

Irino, T. and Patterson, R. D. (**2006**). "A dynamic compressive gammachirp auditory filterbank.", IEEE Trans Audio Speech Lang Processing **14**, 2222–2232.

Jeffress, L. A. (**1948**). "A Place Theory Of Sound Localization", Journal of Comparative and Physiological Psychology **41**, 35–39.

Jeffress, L. A., Blodgett, H. C., Sandel, T. T., and Wood, C. L. (**1956**). "Masking of tonal signals", J. Acoust. Soc. Am. **28**, 416–426.

Jepsen, M. L., Ewert, S. D., and Dau, T. (**2008**). "A computational model of human auditory signal processing and perception", J. Acoust. Soc. Am. **124**, 422–438.

Jiang, D., Palmer, A. R., and Winter, I. M. (**1996**). "Frequency extent of two-tone facilitation in onset units in the ventral cochlear nucleus", J. Neurophysiol. **75**, 380–395.

Johannesen, P. T. and Lopez-Poveda, E. A. (**2008**). "Cochlear nonlinearity in normal-hearing subjects as inferred psychophysically and from distortion-product otoacoustic emissions", J. Acoust. Soc. Am. **124**, 2149–2163.

Kalluri, R. and Shera, C. A. (**2007**). "Comparing stimulus-frequency otoacoustic emissions measured by compression, suppression, and spectral smoothing.", J. Acoust. Soc. Am. **122**, 3562–3575.

Kapfer, C., Seidl, A. H., Schweizer, H., and Grothe, B. (**2002**). "Experience-dependent refinement of inhibitory inputs to auditory coincidence-detector neurons", Nat. Neurosc. **5**, 247–253.

Kemp, D. T. (**1978**). "Stimulated acoustic emissions from within the human auditory system", J. Acoust. Soc. Am. **64**, 1386–1391.

Kemp, D. T. (**1979**). "Evidence of mechanical nonlinearity and frequency selective wave amplification in the cochlea", Archives of oto-rhino-laryngology-Archiv Für Ohren-Nasen- und Kehlkopfheilkunde **224**, 37–45.

Kemp, D. T. and Chum, R. (**1980**). "Properties of the generator of stimulated acoustic emissions.", Hear Res **2**, 213–232.

Kummer, P., Janssen, T., and Arnold, W. (**1998**). "The level and growth behavior of the 2 f1 - f2 distortion product otoacoustic emission and its relationship to auditory sensitivity in normal hearing and cochlear hearing loss", J. Acoust. Soc. Am. **103**, 3431–3444.

Las, L., Stern, E. A., and Nelken, I. (**2005**). "Representation of tone in fluctuating maskers in the ascending auditory system", J. Neurosci. **25**, 1503–1513.

Levitt, H. (**1971**). "Transformed up-down methods in psychoacoustics", J. Acoust. Soc. Am. **49**, 467–477.

Li, H. Z., Sabes, J. H., and Sinex, D. G. (**2006**). "Responses of inferior colliculus neurons to SAM tones located in inhibitory response areas", Hear. Res. **220**, 116–125.

Licklider, J. C. R. (**1948**). "The influence of interaural phase relations upon the masking of speech by white noise", J. Acoust. Soc. Am. **20**, 150–159.

Liu, Q. and Davis, R. L. (**2007**). "Regional specification of threshold sensitivity and response time in cba/caj mouse spiral ganglion neurons", J. Neurophysiol. **98**, 2215–2222.

Long, G. R. and Tubis, A. (**1988**). "Modification of spontaneous and evoked otoacoustic emissions and associated psychoacoustic microstructure by aspirin consumption", J. Acoust. Soc. Am. **84**, 1343–1353.

Lopez-Poveda, E. A. and Alves-Pinto, A. (**2008**). "A variant temporal-masking-curve method for inferring peripheral auditory compression", J. Acoust. Soc. Am. **123**, 1544–1554.

Mauermann, M. and Kollmeier, B. (**2004**). "Distortion product otoacoustic emission (DPOAE) input/output functions and the influence of the second DPOAE source.", J. Acoust. Soc. Am. **116**, 2199–2212.

Mauermann, M., Long, G. R., and Kollmeier, B. (**2004**). "Fine structure of hearing threshold and loudness perception.", J. Acoust. Soc. Am. **116**, 1066–1080.

Mauermann, M., Uppenkamp, S., van Hengel, P. W. J., and Kollmeier, B. (**1999**a). "Evidence for the distortion product frequency place as a source of distribution product otoacoustic emission (DPOAE) fine structure in humans. I. Fine structure and higher-order DPOAE as a function of the frequency ratio f2/f1", J. Acoust. Soc. Am. **106**, 3473–3483.

Mauermann, M., Uppenkamp, S., van Hengel, P. W. J., and Kollmeier, B. (**1999**b). "Evidence for the distortion product frequency place as a source of distortion product otoacoustic emission (DPOAE) fine structure in humans. II. Fine structure for different shapes of cochlear hearing loss", J. Acoust. Soc. Am. **106**, 3484–3491.

McAlpine, D. and Grothe, B. (**2003**). "Sound localization and delay lines - do mammals fit the model?", Trends In Neurosciences **26**, 347–350.

McFadden, D. (**1986**). "Comodulation masking release - effects of varying the level, duration, and time-delay of the cue band", J. Acoust. Soc. Am. **80**, 1658–1667.

McFadden, D. and Wright, B. A. (**1992**). "Temporal decline of masking and comodulation masking release", J. Acoust. Soc. Am. **92**, 144–156.

Meddis, R., Delahaye, R., O'Mard, L., Sumner, C., Fantini, D. A., Winter, I., and Pressnitzer, D. (**2002**). "A model of signal processing in the cochlear nucleus: Comodulation masking release", Acta Acust./Acustica **88**, 387–398.

Meddis, R., O'Mard, L. P., and Lopez-Poveda, E. A. (**2001**). "A computational algorithm for computing nonlinear auditory frequency selectivity", J. Acoust. Soc. Am. **109**, 2852–2861.

Moleti, A., Paternoster, N., Bertaccini, D., Sisto, R., and Sanjust, F. (**2009**). "Otoacoustic emissions in time-domain solutions of nonlinear non-local cochlear models", J. Acoust. Soc. Am. **126**, 2425–2436.

Moore, B. C. J. (**1978**). "Psychophysical tuning curves measured in simultaneous and forward masking", J. Acoust. Soc. Am. **63**, 524–532.

Moore, B. C. J., Hall, J. W., Grose, J. H., and Schooneveldt, G. P. (**1990**). "Some factors affecting the magnitude of comodulation masking release", J. Acoust. Soc. Am. **88**, 1694–1702.

Moore, B. C. J. and Shailer, M. J. (**1991**). "Comodulation masking release as a function of level", J. Acoust. Soc. Am. **90**, 829–835.

Neely, S. T. and Kim, D. O. (**1983**). "An active cochlear model showing sharp tuning and high sensitivity.", Hear Res **9**, 123–130.

Neely, S. T. and Kim, D. O. (**1986**). "A model for active elements in cochlear biomechanics.", J. Acoust. Soc. Am. **79**, 1472–1480.

Nelken, I., Rotman, Y., and Bar Yosef, O. (**1999**). "Responses of auditory-cortex neurons to structural features of natural sounds", Nature **397**, 154–157.

Nelson, D. A., Schroder, A. C., and Wojtczak, M. (**2001**). "A new procedure for measuring peripheral compression in normal-hearing and hearing-impaired listeners", J. Acoust. Soc. Am. **110**, 2045–2064.

Neuert, V., Verhey, J. L., and Winter, I. M. (**2004**). "Responses of dorsal cochlear nucleus neurons to signals in the presence of modulated maskers", J. Neurosci. **24**, 5789–5797.

Nieder, A. and Klump, G. M. (**2001**). "Signal detection in amplitude-modulated maskers. II. Processing in the songbird's auditory forebrain", Eur. J. Neurosci. **13**, 1033–1044.

Nuttall, A. L. and Dolan, D. F. (**1996**). "Steady-state sinusoidal velocity responses of the basilar membrane in guinea pig", J. Acoust. Soc. Am. **99**, 1556–1565.

Ohm, G. (**1843**). "Über die Definition des Tones, nebst daran geknüpfter Theorie der Sirene und ähnlicher tonbildender Vorrichtungen (On the definition of the tone [...])", Ann. Phys. Chem. **59**, 513–565.

Osman, E. (**1971**). "Correlation model of binaural masking level differences", J. Acoust. Soc. Am. **50**, 1494–1511.

Oxenham, A. J. (**2001**). "Forward masking: Adaptation or integration?", J. Acoust. Soc. Am. **109**, 732–741.

Oxenham, A. J. and Plack, C. J. (**1997**). "A behavioral measure of basilar-membrane nonlinearity in listeners with normal and impaired hearing", J. Acoust. Soc. Am. **101**, 3666–3675.

Piechowiak, T., Ewert, S. D., and Dau, T. (**2007**). "Modeling comodulation masking release using an equalization-cancellation mechanism", J. Acoust. Soc. Am. **121**, 2111–2126.

Plack, C. J., Oxenham, A. J., and Drga, V. (**2002**). "Linear and nonlinear processes in temporal masking", Acta Acust./Acustica **88**, 348–358.

Pressnitzer, D., Meddis, R., Delahaye, R., and Winter, I. M. (**2001**). "Physiological correlates of comodulation masking release in the mammalian ventral cochlear nucleus", J. Neurosci. **21**, 6377–6386.

Rhode, W. S. (**1971**). "Observations of vibration of basilar membrane in squirrel monkeys using Mössbauer technique", J. Acoust. Soc. Am. **49**, 1218–1231.

Richards, V. M. (**1987**). "Monaural envelope correlation perception", J. Acoust. Soc. Am. **82**, 1621–1630.

Robles, L. and Ruggero, M. A. (**2001**). "Mechanics of the mammalian cochlea.", Physiol Rev **81**, 1305–1352.

Ruggero, M. A., Rich, N. C., Recio, A., Narayan, S. S., and Robles, L. (**1997**). "Basilar-membrane responses to tones at the base of the chinchilla cochlea", J. Acoust. Soc. Am. **101**, 2151–2163.

Ruggero, M. A., Robles, L., and Rich, N. C. (**1992**). "2-Tone suppression in the basilar-membrane of the cochlea - Mechanical basis of auditory-nerve rate suppression", J. Neurophysiol. **68**, 1087–1099.

Russell, I., Cody, A., and Richardson, G. (**1986**). "The responses of inner and outer hair cells in the basal turn of the guinea-pig cochlea and in the mouse cochlea grown in vitro", Hearing Research **22**, 199 – 216.

Schloth, E. and Zwicker, E. (**1983**). "Mechanical and acoustical influences on spontaneous oto-acoustic emissions", Hear. Res. **11**, 285 – 293.

Schooneveldt, G. P. and Moore, B. C. J. (**1987**). "Comodulation masking release (CMR) - Effects of signal frequency, flanking-band frequency, masker bandwidth, flanking-band level, and monotic versus dichotic presentation of the flanking band", J. Acoust. Soc. Am. **82**, 1944–1956.

Schooneveldt, G. P. and Moore, B. C. J. (**1989**). "Comodulation masking release for various monaural and binaural combinations of the signal, on-frequency, and flanking bands", J. Acoust. Soc. Am. **85**, 262–272.

Shera, C. A. (**2003**). "Mammalian spontaneous otoacoustic emissions are amplitude-stabilized cochlear standing waves", J. Acoust. Soc. Am. **114**, 244–262.

Shera, C. A. and Zweig, G. (**1993**). "Noninvasive measurement of the cochlear traveling-wave ratio", J. Acoust. Soc. Am. **93**, 3333–3352.

Strickland, E. A. and Viemeister, N. F. (**1996**). "Cues for discrimination of envelopes", J. Acoust. Soc. Am. **99**, 3638–3646.

Talmadge, C. L., Tubis, A., Long, G. R., and Piskorski, P. (**1998**). "Modeling otoacoustic emission and hearing threshold fine structures", J. Acoust. Soc. Am. **104**, 1517–1543.

Thomas, I. B. (**1975**). "Microstructure of the pure tone threshold", J. Acoust. Soc. Am. Suppl. 1 **75**, S26–S27.

Thompson, S. K., von Kriegstein, K., Deane-Pratt, A., Marquardt, T., Deichmann, R., Griffiths, T. D., and McAlpine, D. (**2006**). "Representation of interaural time delay in the human auditory midbrain", Nat. Neurosc. **9**, 1096–1098.

v. Bekesy, G. (**1949**a). "On the Resonance Curve and the Decay Period at Various Points on the Cochlear Partition", J. Acoust. Soc. Am. **21**, 245–254.

v. Bekesy, G. (**1949**b). "The Vibration of the Cochlear Partition in Anatomical Preparations and in Models of the Inner Ear", J. Acoust. Soc. Am. **21**, 233–245.

van de Par, S. and Kohlrausch, A. (**1997**). "A new approach to comparing binaural masking level differences at low and high frequencies", J. Acoust. Soc. Am. **101**, 1671–1680.

van de Par, S. and Kohlrausch, A. (**1999**). "Dependence of binaural masking level differences on center frequency, masker bandwidth, and interaural parameters", J. Acoust. Soc. Am. **106**, 1940–1947.

van den Raadt, M. P. M. G. and Duifhuis, H. (**1990**). *The Mechanics and Biophysics of Hearing* (Springer-Verlag, Berlin).

van der Heijden, M. and Trahiotis, C. (**1997**). "A new way to account for binaural detection as a function of interaural noise correlation", J. Acoust. Soc. Am. **101**, 1019–1022.

van Hengel, P. W., Duifhuis, H., and van den Raadt, M. P. (**1996**). "Spatial periodicity in the cochlea: the result of interaction of spontaneous emissions?", J. Acoust. Soc. Am. **99**, 3566–3571.

Verhey, J. L., Dau, T., and Kollmeier, B. (**1999**). "Within-channel cues in comodulation masking release (CMR): Experiments and model predictions using a modulation-filterbank model", J. Acoust. Soc. Am. **106**, 2733–2745.

Verhey, J. L., Pressnitzer, D., and Winter, I. M. (**2003**). "The psychophysics and physiology of comodulation masking release", Exp. Brain Res. **153**, 405–417.

Verhey, J. L., Rennies, J., and Ernst, S. M. A. (**2007**). "Influence of envelope distributions on signal detection", Acta Acust./Acustica **93**, 115–121.

Viemeister, N. F. (**1979**). "Temporal-modulation transfer-functions based upon modulation thresholds", J. Acoust. Soc. Am. **66**, 1364–1380.

Winter, I. M. and Palmer, A. R. (**1995**). "Level dependence of cochlear nucleus onset unit responses and facilitation by 2nd tones or broad-band noise", J. Neurophysiol. **73**, 141–159.

Zurek, P. M. and Durlach, N. I. (**1987**). "Masker-bandwidth dependence in homophasic and antiphasic tone detection", J. Acoust. Soc. Am. **81**, 459–464.

Zhou, Z. P., Liu, Q., and Davis, R. L. (**2005**). "Complex regulation of spiral ganglion neuron firing patterns by neurotrophin-3", J. Neurosci. **25**, 7558–7566.

Zweig, G. (**1991**). "Finding the impedance of the organ of corti", J. Acoust. Soc. Am. **89**, 1229–1254.

Zweig, G., Lipes, R., and Pierce, J. R. (**1976**). "The cochlear compromise", J. Acoust. Soc. Am. **59**, 975–982.

Zweig, G. and Shera, C. A. (**1995**). "The origin of periodicity in the spectrum of evoked otoacoustic emissions", J. Acoust. Soc. Am. **98**, 2018–2047.

Zwicker, E. and Schloth, E. (**1984**). "Interrelation of different oto-acoustic emissions.", J. Acoust. Soc. Am. **75**, 1148–1154.

List of Figures

Figure 2.1 : Schematic spectrogram of the uncorrelated (UN) and comodulated (CM) masker conditions 24
Figure 2.2 : Envelope amplitude statistics of multiplied and Gaussian noise 29
Figure 2.3 : Schematic structure of the model as a hypothetical neural circuit and the functional implementation 36
Figure 2.4 : Internal representations within the model for uncorrelated and comodulated stimulus conditions 40
Figure 2.5 : Model predictions (right panels) adjusted to experimentally obtained data (left panels) 42
Figure 2.6 : Data and simulations for multiplied noise maskers in uncorrelated (UN) and comodulated (CM) conditions 43
Figure 2.7 : Same as Figure 2.6 for Gaussian noise maskers. 45
Figure 2.8 : Data and simulations for different portions of comodulated masker 48

Figure 3.1 : Individual data for a signal frequency of 700 Hz and the multiplied noise masker . 67
Figure 3.2 : Mean results for the multiplied noise masker 69
Figure 3.3 : As Fig. 3.2 but for the Gaussian noise masker. 71

Figure 3.4 : Results for a signal frequency of 200 Hz with multiplied and Gaussian noise maskers. 74
Figure 3.5 : As Fig. 3.4, but for a signal frequency of 3000 Hz. 76
Figure 3.6 : Time signal and spectrum of a Gaussian noise masker sample for a transposed stimulus 79
Figure 3.7 : As Fig. 3.5, but for the transposed stimulus. 80

Figure 4.1 : Damping profile and feedback stiffness of the active and nonlinear transmission line model of the cochlea 102
Figure 4.2 : Excitation patterns for a 1 kHz pure tone with levels from 0-120 dB SPL. 107
Figure 4.3 : Input-output curve at the CP of a 1-kHz pure tone with a level of 0 dB SPL. 109
Figure 4.4 : Two-tone suppression map for a suppressee frequency of 1 kHz and 40 dB SPL. 112
Figure 4.5 : Tonic suppression for a suppressee of 1 kHz and a suppressor of 500 Hz. 114
Figure 4.6 : Level of the $2f_1 - f_2$ DPOAE distortion component (wave-fixed component only). 116
Figure 4.7 : Magnitude and phase of SFOAE 119
Figure 4.8 : Excitation patterns of the model with roughness (solid lines) and without roughness (dashed line) 122
Figure 4.9 : Simulated threshold in quiet and SOAE. . 123
Figure 4.10 : Modulation detection thresholds in dB modulation depth. Individual data and simulations. 126
Figure A.11 : Excitation patterns of two single-sigmoidal damping functions and the double sigmoidal damping function used in the present study. 136

Figure B.12 : Simulated threshold in quiet for different values of the roughness scaling coefficient (parameter ϵ in Tab. 4.1) 138

List of Tables

Table 4.1 : Parameters used for the simulations. The parameters were fixed for all simulations. 104

Danksagung

Im Laufe dieser Arbeit konnte ich von der Hilfsbereitschaft, der Diskussionsfreudigkeit, dem Verständnis und der Förderung vieler Menschen profitieren. Dieses konnte ich sowohl im univeristären als auch im privaten Umfeld spüren, wodurch mir viele Dinge ermöglicht wurden. Dafür möchte ich meinen Dank aussprechen.

Finanziell wurde ich im Laufe dieser Arbeit unterstützt durch Stipendien der Deutschen Forschungsgemeinschaft (DFG) im Rahmen des „Internationalen Graduiertenkollegs Neurosensorik“ (GRK 591), des SFB „Aktives Gehör“ (SFB TR31) und einem Doktorandenstipendium der Heinz Neumüller Stiftung. Auch hierfür möchte ich mich bedanken.

Lebenslauf

Bastian Epp

geboren am 22. Dezember 1981
in Heilbronn

Staatsangehörigkeit: deutsch

Feb. 2007 bis Mai 2010	Promotion in der Arbeitsgruppe „Neurosensorik" (Prof. Dr. J. L. Verhey) an der Carl-von-Ossietzky Universität Oldenburg
Nov. 2006	„Master of Science (Eng. Phys.)" zum Thema „Interaction of monaural across-frequency and binaural processing" (Prof. Dr. J. L. Verhey)
Apr. 2005	„Bachelor of Engineering (Eng. Phys.)" zum Thema „Investigation of a parametric sound source for audio sound" (Prof. Dr. V. Mellert)
Okt. 2002 bis Nov. 2006	Studium „Engineering Physics" an der Carl-von-Ossietzky Universität Oldenburg und an der Danmarks Tekniske Universitet (DTU) in Kopenhagen, Dänemark
Okt. 2001 bis Jun. 2002	Zivildienst
Apr. 2001	Abitur am Ganztagesgymnasium Osterburken